GLOBALIZING
CUSTOMER SOLUTIONS

GLOBALIZING CUSTOMER SOLUTIONS

The Enlightened Confluence
of Technology, Innovation,
Trade, and Investment

EDMUND B. FITZGERALD

Westport, Connecticut
London

HC
79
.T4
F57
2000

Library of Congress Cataloging-in-Publication Data

Fitzgerald, Edmund B., 1926–
    Globalizing customer solutions : the enlightened confluence of technology,
    innovation, trade, and investment / Edmund B. Fitzgerald.
        p.   cm.
    Includes bibliographical references and index.
    ISBN 0–275–96995–9 (alk. paper)—ISBN 0–275–97073–6 (pbk : alk. paper)
    1. Technological innovations—Economic aspects.   2. Information
technology—Economic aspects.   3. International trade.   4. Investments, Foreign.   5.
Competition, International.   I. Title.
HC79.T4F57   2000
382—dc21   00–025462

British Library Cataloguing in Publication Data is available.

Library of Congress Catalog Card Number: 00–025462
ISBN: 0–275–96995–9
       0–275–97073–6 (pbk.)

First published in 2000

Praeger Publishers, 88 Post Road West, Westport, CT 06881
An imprint of Greenwood Publishing Group, Inc.
www.praeger.com

Printed in the United States of America

The paper used in this book complies with the
Permanent Paper Standard issued by the National
Information Standards Organization (Z39.48–1984).

10 9 8 7 6 5 4 3 2 1

Contents

Preface

I believe that it is reasonable to ask, Why write a book about globalizing customer solutions and the tactics of technology, innovation, international trade, and foreign direct investment that underlie it? Simply stated, such globalization is conducive to the further deployment of the twin ideologies of political democracy and the market force economics so important to sustained improvement in global economic welfare.

In a business career spanning more than fifty years, I have been privileged to enjoy a close relationship with the globalization of customer solutions, which has caused me to postulate that the enlightened confluence of technology, innovation, international trade, and foreign direct investment is, indeed, the key ingredient to superior levels of global competitiveness. Or, said another way, these elements are the resources from which "conspicuous customer solutions" are created.

The United States is the acknowledged global leader in generating new technology and adopting this new knowledge to innovative new products, processes, and/or services. However, I have often been surprised to observe how less knowledgeable and often less interested many Americans seem to be in the complementary globalization tactics of international trade and foreign direct investment as contrasted to their counterparts in other nation states. The massive and homogeneous U.S. domestic market creates economic opportunities that are the envy of the world. However, this massive domestic U.S. market also often diminishes the incentives for domestic enterprises to actively pursue opportunities beyond the boundaries of the United States and/or NAFTA (North American Free Trade Agreement).

I wish to dedicate this book to the U.S. Committee for Economic Development (CED), an independent research and policy organization of some 250 business leaders and educators. The CED is nonprofit, nonpartisan, and nonpolitical. Its purpose is to propose policies that support steady economic growth at high employment and reasonably steady prices, which support increased productivity and living standards, greater and more equal opportunity for every citizen, and an improved quality of life for all. It has been my good fortune to have served as a trustee of the CED for over thirty years and as chairman of the CED's board of trustees from 1982 until 1986.

More specifically, I wish to dedicate this book to Dr. Isaiah Frank, the William L. Clayton Professor of International Economics at the Johns Hopkins University School of Advanced International Studies (SAIS) in Washington, D.C., who for most of my tenure with the CED has served as the adviser on international economic policy. Although I have been deeply involved in both technology and innovation since my university days, my enhanced interest in and knowledge of international trade and foreign direct investment is directly due to the many experiences that Dr. Frank and I have shared in the development of CED policy statements on these subjects.

This work is liberally sprinkled with "the gospel according to Isaiah," in both a conscious and unconscious fashion. I hope Isaiah will be pleased that one of his collaborators listened so intently to his wise observations, but any errors or shortcomings in this book are mine alone and should not be attributed to him. Having spent a portion of my business career in the telecommunications industry, one of my favorite Isaiah Frank quotes is his definition of an economist as "someone who, if you forget your telephone number, will offer to estimate it for you."

Finally, I would like to thank the late Dr. Martin S. Geisel, former dean of the Owen Graduate School of Management at Vanderbilt University, in Nashville, Tennessee, who during the 1990s graciously permitted me, as an adjunct professor, to explore the subjects of technology, innovation, international trade, and foreign direct investment with the master's of business administration (MBA) and executive MBA candidates at the Owen School. In view of my concerns regarding insufficient interest in and knowledge of trade and foreign investment among contemporary Americans, this has been a rare and satisfying opportunity to gain greater exposure of these important subjects to so many bright, energetic young minds.

Abbreviations

APEC	Asia/Pacific Economic Cooperation
CED	Committee for Economic Development
CEO	Chief Executive Officer
COCOM	Coordinating Committee of Multilateral Export Controls
COMECON	Council for Mutual Economic Cooperation
DJIA	Dow Jones Industrial Average
DSB	Dispute-Settlement Body
EU	European Union
FCPA	Foreign Corrupt Practices Act
FDI	Foreign Direct Investment
FTA	Free Trade Agreement
FTAA	Free Trade Area for the Americas
GATT	General Agreement on Tariffs and Trade
GDP	Gross Domestic Product
IBR&D	International Bank for Reconstruction & Development
ICCCA	International Chamber of Commerce Court of Arbitration

ICSID	International Center for the Settlement of Investment Disputes
IMF	International Monetary Fund
IPR	Intellectual Property Rights
IRS	Internal Revenue Service
ISO	International Standards Organization
ITC	International Trade Commission
ITO	International Trade Organization
MAI	Multilateral Agreement on Investment
MBA	Master of Business Administration
MFA	Multifibre Agreement
MFN	Most Favored Nation
MIGA	Multilateral Investment Guarantee Agency
MITI	Ministry of International Trade and Investment
MNC	Multinational Corporation
NAFTA	North American Free Trade Agreement
NGO	Non-Governmental Organization
NIC	Newly Industrialized Countries
NIEO	New International Economic Order
ODA	Official Development Assistance
OECD	Organization for Economic Cooperation & Development
OMA	Orderly Marketing Agreement
OPEC	Organization of Petroleum Exporting Countries
OPIC	Overseas Private Investment Corporation
PAC	Political Action Committee
PEC	President's Export Council
P&L	Profit and Loss
PPP	Purchasing Power Parity
R&D	Research & Development
SAIS	School of Advanced International Studies (Johns Hopkins)
TRIMS	Trade-Related Investment Measures
TRIPS	Trade-Related Intellectual Property Rights
TNC	Transnational Corporation

UNCTAD	United Nations Commission on Trade and Development
VAT	Value-Added Tax
VERs	Voluntary Export Restraints
VRAs	Voluntary Restraint Agreements
WEF	World Economic Forum
WHFTA	Western Hemisphere Free Trade Area
WTO	World Trade Organization

Chapter 1

Introduction and the Global Economic Metrics

The close relationship between technology and innovation has long been recognized, in that innovation is the process by which the value inherent in technology is realized. Similarly, the close relationship between international trade and foreign direct investment has long been recognized, in that they are complementary business tactics by which private-sector enterprises seek to optimize the global deployment of their resources.

Less obvious has been the close relationship of technology and innovation to trade and investment, through the latter's role of significantly expanding the dimension of economic opportunity inherent in technology and innovation by globalizing the addressable market for that innovation in an era of rapidly decreasing product life cycles. As the addressable market is enlarged, so is the ability to more rapidly and economically amortize the significant research-and-development (R&D) expenses incurred in producing today's new innovative products and/or services. Currently, in many product sectors, the R&D amortization cost per unit exceeds the unit's production cost. The above close relationships establish the foundation for my thesis, that the enlightened confluence of technology, innovation, trade, and investment is, indeed, the critical issue in creating "conspicuous customer solutions," which I define as customized, highest-value products and/or services of zero-deficit quality, at least cost and created in least elapsed time.

Although not specifically addressed in this book, it is obvious that technology, innovation, trade, and investment best flourish in an environment of political and economic freedom. Political democracy can't be sacrificed for economic development. Policy makers and investors around the globe increasingly understand that to be free is to grow and to prosper. There is an important role for national governments and international institutions in protecting free markets, principally through their judicial, monetary, and international functions. However, it is increasingly apparent that the most important factors in determining long-term economic growth and prosperity are policies that reduce, to the greatest extent possible, government constraints on economic activity. A potential danger from the 1997–98 economic turmoil in Asia is that national leaders will mistakenly conclude that their economies need more protection from the global marketplace.

THE METRICS OF THE WORLD ECONOMY

It is estimated by the International Monetary Fund (IMF) and the World Bank that the world will exit the twentieth century with an aggregated annual gross product output of just over U.S. $31 trillion, based on currency-market exchange rates. World gross product output aggregates the gross product or value added of all world nation's as distributed in the form of compensation (the income of employees), profits (the income of enterprises), and taxes (the income of governments).

Economists, who never like to make matters simple, prefer to aggregate world output on the basis of purchasing power parity (PPP). Under the PPP technique, the cost of a basket of life's necessities is calculated in every nation and then compared against the cost for this same basket in the United States. The gross domestic product (GDP) of nations with costs higher than the U.S. cost are then reduced by the percentage of their cost excesses, and the gross domestic products of nations with costs lower than the U.S. cost are then increased by the percentages that their costs are less than the U.S. model.

Even though high-cost developed nations, such as Luxembourg,

Switzerland, and Japan experience a 25% to 40% reduction in gross domestic product under the PPP calculation, the major impact of this exercise is a significant increase in the gross domestic product of the nations of the developing world, whose share of world output is virtually doubled through the PPP calculation. In general, services are the primary undervalued commodity in the developing world. When adjusted for PPP, estimated world output in the year 2000 rises from just over U.S.$31 trillion to just over U.S.$43 trillion.

Currently, the World Bank publishes economic data on 210 nations, and the IMF publishes similar data on 184 nations. According to the World Bank, the global economy will exit the twentieth century with a population of about six billion people. If global output were equally distributed among the world's population, the global output per capita, on a market exchange rate basis, would be just over U.S.$5,000. However, since world output is not evenly distributed, national output per capita becomes the best measure by which to compare living standards in individual nations. Output per capita determines potential compensation per capita, and it is compensation per capita that determines relative living standards. Recent World Bank data indicates that on a market exchange rate basis, approximately 80% of the world's annual output is produced by a group of nations in which only about 15% of the world's population resides, and that only 12% of the world's annual output is produced by a group of nations in which over 75% of the world's population resides.

THE IMF CLASSIFICATION OF NATIONS

As noted earlier, the IMF publishes economic data on 184 nations. As shown in Figure 1.1, the IMF classifies these 184 nations as being either advanced economies (28), countries in transition (28), or developing nations (128).

The IMF advanced economies are nations of significant economic similarity. Most are members of the Organization for Economic Cooperation and Development (OECD) in Paris. Fifteen of the twenty-eight are member states of the European Union. All but Australia and New Zealand are in the Northern Hemisphere. On a PPP basis, the average annual GDP per capita within these

Figure 1.1
The IMF Classification of Nations

<u>**Advanced Economies**</u>

Australia	Hong Kong, SAR	Norway
Austria*	Iceland	Portugal*
Belgium*	Ireland*	Singapore
Canada	Israel	South Korea
Denmark*	Italy*	Spain*
Finland*	Japan	Sweden*
France*	Luxembourg*	Switzerland
Germany*	Netherlands*	Taiwan
Greece*	New Zealand	United Kingdom*
		United States

*Members of the European Union (15)

<u>**Countries in Transition**</u> from a system of centralized administrative control to one based on market principles.

Albania	Hungary	Poland
Armenia	Kazakhstan	Romania
Azerbaijan	Kyrgyz Republic	Russia
Belarus	Latvia	Slovak Republic
Bosnia & Herzegovina	Lithuania	Slovenia
Bulgaria	Macedonia, Former	Tajikistan
Croatia	Rep. Of Yugoslavia	Turkmenistan
Czech Republic	Moldova	Ukraine
Estonia	Mongolia	Uzbekistan
Georgia		Yugoslavia, Federal Republic of Serbia/Montenegro

<u>**Developing Countries**</u>— All other nations except a few, such as Cuba and the DPRK (North Korea), which are not IMF members, or because databases have not yet been compiled.

nations is about U.S.$25,000. On a PPP basis, their individual GDP per capita's range from a low of approximately U.S.$13,000 to a high of approximately U.S.$34,000.

The similarity between the twenty-eight countries in transition is that they are transitioning from a socialist, centrally administered economic model toward more market-based economies. Many of these nations are former members of the Council for Mutual Economic Cooperation (COMECON), the former Soviet–Eastern European economic union, which was disbanded at the end of the cold war.

The remaining IMF classification is developing nations, of which there are 128. These nations have very little similarity between themselves. A great many of them are located in the Southern Hemisphere. Economic similarity is not a characteristic of developing nations, nor of countries in transition. A nation such as Slovenia has a GDP per capita comparable to that of the advanced economies. A nation such as Ethiopia has a GDP per capita of only U.S.$100. The vast economic, social, and political differences between the nations in these latter two IMF categories would certainly suggest that in such divergent circumstances, one economic or commercial strategy is not likely to fit all.

IMF estimates for world output shares in the year 2000 and the annual output growth rates during the decade of the 1990s for the world, selected regional areas, and selected nations are shown in Figure 1.2.

INTERNATIONAL TRADE

Obviously a nation's economic output is not totally consumed domestically. A portion of domestic output is exported to other nations, and portions of domestic output from other nations are imported. By the end of the twentieth century, according to IMF estimates, annual global exports of both goods and services should total U.S.$7.5 trillion. Trade in goods should constitute 80% of total international trade, and trade in services should account for 20%. Advanced economies are expected to account for 78% of these exports, countries in transition 4%, and developing nations about 18%. During the 1990s, annual exports of goods and services are expected to have grown more than twice as fast as world output—6.2% for exports, compared to 3.2% for world output growth.

A NATION'S CURRENT ACCOUNT BALANCE

On a global basis, international trade is symmetrical, that is to say, for each export there must be an import, and vice versa. However, on a national basis, trade is seldom symmetrical. A nation's exports seldom match its imports, which creates what economists call national current account balances. A nation's current account balance consists of the sum of:

Figure 1.2
World Output Shares in 2000 (on PPP basis)

	Percentage Share	Annual Growth Rate 1990s
World	100.0%	3.2%
Advanced Economies	55.4	2.4
United States	20.8	2.7
European Union	19.9	2.0
Japan	7.4	1.3
Developing Countries	39.8	5.5
Asian Nations	22.8	7.3
China	12.0	10.7
Countries in Transition	4.8	(3.2)
Russia	1.6	(5.7)

Source: IMF World Economic Outlook, October 1999

- *Merchandise trade balance*—the difference between the imports and exports of goods (agricultural and forest products raw materials, fuels, manufactured parts, and finished products).
- *Services trade balance*—the difference between the imports and exports of services such as transportation, banking, insurance, construction, and communications, plus *net* investment income.
- *Unilateral transfers*—pensions, remittances, and other official and unofficial transfers.

It is important to remember that a nation's current account balance excludes capital transactions. Ordinarily, the IMF's advanced nations, as a group, enjoy current account surpluses, and developing nations, plus countries in transition, as a group, experience current account deficits. However, for almost twenty years the United States has suffered substantial current account deficits on a global basis, on a bilateral basis with Japan, and, in recent years, on a bilateral basis with China. The primary U.S. deficit category has been merchandise trade, which is significantly influenced by massive net imports of automobiles, auto parts, and foreign oil. The United States has a significant and growing surplus in services trade, in which it enjoys a substantial comparative advantage. But at the same time, the United States has experienced a declining and now negative balance in net investment income. The U.S. unilateral transfers are usually around U.S.$40 billion in deficit, but they experienced a small surplus in 1991, when the U.S. allies in the Gulf War reimbursed the United States for military expenses incurred during the war.

For many years, the large U.S. merchandise trade deficit has been a major U.S. political issue. Many claim that liberal trade policies have not well served the U.S. interest, and that U.S. trading partners do not abide by the same trade rules as does the United States. There is no doubt that unfair trade practices by certain U.S. trading partners have done some harm to U.S. trade balances, but the total impact of such practices has probably accounted for considerably less than 10% of past U.S. trade deficits.

Primary culprits have been U.S. domestic policy instruments, which favored domestic consumption over domestic savings, including a more than twenty-year history of large federal budget deficits. These proconsumption domestic policies have often juxtaposed policies of economic constraint in the other nation states with whom the U.S. trades, and have created increased U.S. demand for imports and reduced demand for U.S. exports. During this same period, significant progress by the General Agreement on Tariffs and Trade (GATT) and the World Trade Organization (WTO) in reducing trade barriers and eliminating unfair trade practices has reduced or eliminated many of these practices that the United States earlier perceived on the part of its trading partners, especially Japan. Changes in global currency exchange rates will always influence trade balances, most probably to the detriment of the United States, when financial crises erupt in other global nations and/or regions.

A NATION'S CAPITAL ACCOUNT BALANCE

As noted earlier, a nation's capital account transactions are excluded from a nation's current account balance. A nation's capital account interacts with a nation's current account in a similar fashion to that in which a private enterprise's capital transactions interact with its profit-and-loss (P&L) statement. If a private enterprise's P&L statement is in deficit and that enterprise needs additional liquidity, it either borrows funds or sells assets. Similarly, if its P&L is in surplus and it is accumulating cash, that enterprise frequently seeks new investment opportunities. The experience of nations is similar. Thus, a nation's capital account balance should be equal and opposite to its current account balance. Statistical discrepancies in the collection of data often cause these two accounts to appear not to offset, but these statistical discrepancies are a small fraction of the sums involved in calculating these balances.

Thus, at the macroeconomic level, a nation's net international flow of capital is simply the counterpart of that nation's imbalance in its current account. A current-account surplus nation is a net capital exporter; a current-account deficit nation is a net capital importer.

As a nation with an endemic current-account deficit, the United States is, therefore, an endemic capital importer. And as long as the sum of domestic public and private expenditures continues to exceed gross domestic product, a nation's current-account balance will remain in deficit and that nation will continue to be a net importer of foreign capital.

Since 1982, the U.S. current account has been in deficit, meaning that during this period U.S. combined public and private expenditures have exceeded the U.S. gross domestic product. In the decade of the 1990s alone, the United States will have imported over U.S.$1.3 trillion in capital to offset its annual current account deficits. During much of this time, the U.S. federal budget was deeply in deficit, adding to total U.S. expenditures. In 1998 and 1999, the U.S. government achieved its first budget surpluses in over twenty years.

Foreign capital is not to be feared, except if its magnitude and liquidity reaches the level at which its precipitous withdrawal could trigger a domestic capital crisis. However, the rents on foreign capital accrue to the foreign investor and provide foreign, not U.S., economic growth. Capital rents accruing to foreign investors represent a net transfer of wealth from the United States to other nations. The United States urgently needs to review the long-term consequences of its high-consumption, low-savings lifestyle, particularly as it impacts or is impacted by rapidly changing domestic demographic patterns.

FOREIGN DIRECT INVESTMENT (FDI)

Imbedded in global financial flows are foreign direct investment flows. Simply defined, foreign direct investment is a cross-border, private sector flow of equity or loan capital, through which the investor intends to realize some form of control over the acquired asset and which is normally a less liquid and longer-term commitment than envisioned in a typical portfolio investment. Ordinarily, the annual inward FDI flows to the IMF advanced economies are smaller than advanced economies' outward FDI flows, the reverse being true for developing nations and countries in transition. However, in recent years, cross-border mega-mergers between global enterprises based in advanced economies have both accelerated the growth and distorted traditional patterns of FDI inflows and outflows. Typical of these cross-border mega-mergers have been those involving British Petroleum/Amoco, U.S. West/Global Crossing, Daimler/Chrysler, and Deutsche Bank/Banker's Trust.

As shown in Figure 1.3, the United Nation's Commission on Trade and Development (UNCTAD) estimates that in 1999 global FDI flows rose to U.S.$800 billion, up 72% from 1997, and FDI inflows into advanced economies rose to U.S.$640 billion, a 134% increase from 1997. At the same time, the level of FDI inflows into the developing world continued to decline, primarily as an aftermath of the Asian financial crisis of 1997–98. Typically, annual FDI flows account for about 8% of global fixed capital formation, for about 6% in the advanced economies and 10% in the developing world. The ratio of FDI inward stocks to GDP is 12% on a global basis, 10% in the advanced nations, and almost 17% in the developing world. In China, this latter ratio is almost 25%.

By the late 1990s, the global sales revenues of foreign direct investment affiliates were estimated to have risen to about U.S. $11.5 trillion, and their gross product (value added) to have reached almost U.S.$2.7 trillion, or 9% of global economic output. The global sales revenues of these FDI affiliates then exceeded the annual exports of goods and services by a factor of 1.7 to 1, thus

Figure 1.3
Global Foreign Direct Investment Flows (in billions of U.S. dollars)

	FDI Inflows			FDI Outflows		
	1997	1998	1999(E)	1997	1998	1999 (E)
Total World	$464	$644	$800	$475	$649	$800
Advanced Nations	$273	$460	$640	$407	$595	$760
EU	126	230		218	386	
U.S.	109	193		110	183	
Japan	3	3		26	24	
Developing Nations	$173	$166	$160	$ 65	$ 52	$ 40
Africa	8	8		1	0	
Latin America	68	72		16	15	
Asia	96	86		48	36	
China	42	44		3	2	
Central and Eastern Europe	$ 18	$ 18		$ 3	$ 2	

Source: United Nations, *World Investment Report*, 1999

establishing international production as the primary source for reaching global markets.

However, it must be remembered that many of these FDI affiliates not only depend on foreign trade for many of their inputs, but also account for U.S.$2.4 trillion in exports. Nevertheless, in the 1990s FDI grew at a rate more than twice as fast as international trade and more than four times as fast as global economic output.

Chapter 2

Globalization: The New Paradigm

Webster's Collegiate Dictionary defines a paradigm as an outstandingly clear or typical example or pattern. As used in the context of this chapter, "pattern" would appear to be the most appropriate definition. Those that laud the emergence of the globalization paradigm point to the faster growth of international trade, compared to world output, and to the increased flows of both portfolio and foreign direct investment. Consumers now choose from a wider variety of foreign as well as domestic goods, and individual savers now regularly invest in diversified portfolios of foreign equities and securities.

The dynamics of specialization, competition, and economic adjustment, inherent in the process of international exchange, have greatly benefited global economic welfare, as has the funding of attractive foreign investment with capital from nations whose stocks of investable funds exceed the availability of favorable domestic investment opportunities.

Critics of this new paradigm argue that globalization creates increased competition from low-wage nations that could destroy developed world jobs by pushing down wages and reducing labor standards. Other concerns are the export of inadequate environmental standards and greater incentives for "runaway plants." As the world globalizes, sovereign national governments have less latitude to set their own economic agendas, and as the Asian financial crisis in 1997–98 demonstrated, there is a downward side to unfettered international capital flows.

MERCANTILISM VERSUS GLOBALIZATION

Other critics of the globalization paradigm have been known to compare it to the economic concept of mercantilism practiced in centuries past. Again, referring to *Webster's Collegiate Dictionary*, *mercantilism* is defined as "an economic system developing during the decay of feudalism to unify and increase the power and especially the monetary wealth of the nation by a strict governmental regulation of the entire national economy usually through policies designed to secure an accumulation of bullion, a favorable balance of trade, the development of agriculture and manufactures, and the establishment of foreign trading monopolies."

It could be said that the globalization paradigm shares certain goals with mercantilism, such as improved economic welfare, trade expansion, and the development of agricultural and manufacturing assets. But the similarity stops there. Mercantilism sought its objectives through strict government regulation of an entire economy. Today's global paradigm is based on an extraordinary worldwide shift in both ideology and policy toward market-driven rather than government-directed solutions to national economic problems. Nor does the current paradigm represent the beggar-thy-neighbor attitude of "I've got mine, how are you doing?" so typical of mercantilism. The international institutions created since the end of World War II have been specifically designed to extend the global benefits of both finance and trade to all world nations.

GLOBALIZATION NOW AND THEN

Many would have us believe that this new globalization paradigm is unique in history. However, this paradigm is paralleled by a similar degree of world economic integration experienced in the early years of the twentieth century, as shown on Figure 2.1. Only in recent years have four of the five developed nations achieved higher ratios of merchandise trade to GDP than they

Figure 2.1
Merchandise Trade Trends

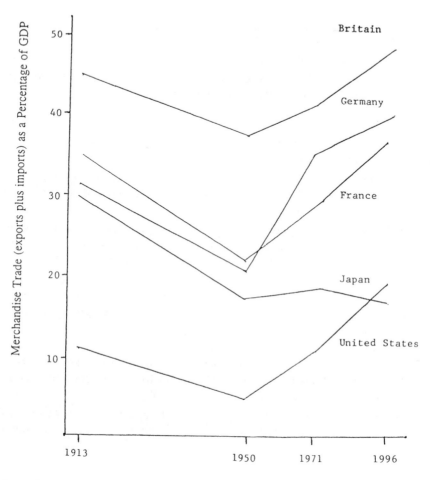

Source: International Monetary Fund and Organization for Economic Cooperation and Development

enjoyed in 1913. The fifth nation, Japan, has never regained its 1913 ratio. Not unlike the 1990s paradigm, this prior globalization era was driven by a reduction in trade barriers and transportation costs. The era ended in 1914 with the outbreak of Word War I, followed by the Great Depression, and then World War II, during which the global economy suffered severe dislocations, world trade decreased, and global economic integration atrophied. The details of this particular era, 1914–50, are more fully discussed in Chapter 4, which briefly traces the history of international trade.

ELEMENTS FAVORING GLOBALIZATION

The globalization paradigm of the late twentieth century had its origins in both technological and regulatory change. Specifically, it was driven by these important events:

- Vastly improved and significantly lower-cost transportation, communication, and information systems.
- Accelerated technological change and the need for more and larger markets to more efficiently amortize higher research-and-development costs over shortened product lives.
- Improved global intellectual property right protection, as achieved in the GATT Uruguay Round, that encourages investment in the high-cost and high-risk process of research and development.
- Significant reductions in import tariffs and other trade barriers. According to IMF staff estimates, post–Uruguay Round average developed-nation tariffs have fallen to 4%, and developing-nation average tariffs to 14%.
- Through the efforts of GATT and the new WTO, restrictions on foreign direct investment have been reduced, more liberal right-of-establishment laws have been enacted, and there is closer supervision of trade-related investment measures (TRIMS).
- Through the processes of deregulation and privatization, more intense domestic competition within more open domestic markets has been achieved.

Few global economic sectors have been impacted more by technological change and innovation than the information industry. Digital technologies and high-bit-rate transmission capabilities have significantly reduced the cost, and increased the speed and capacity, of both communication and information systems. Ultra-wide bandwidth terrestrial and submarine cable systems now circle the globe, providing high-capacity, low-cost information pipelines, and are paralleled by multi-transponder satellite systems. Wireless, cellular, and personal communication systems provide low-cost roving access to individual subscribers and the portability of subscriber identification.

Low-cost, long-haul, and wide-bandwidth transmission channels have reduced the need for on-site, face-to-face meetings and, in many cases, such physical meetings are being replaced by videoconferencing. Time-zone incompatibility has been overcome by fax and E-mail. This improved system of direct, electronic communication has also promoted English as the international language of commerce, a role Esperanto tried and failed to achieve in the years prior to World War II.

The significant impact that these new technologies have had on voice communication pales in comparison to its impact on data communication, which in recent years has been growing at ten times the rate of voice. Major improvements in computer technology; lower-cost, massively increased computing power and speed; and distributed computer engines, including desk-top and lap-top units, have all added to the further integration of both local and wide-area computer networks. The rapid growth of the Internet and Internet protocol have vastly increased the use of packet-and-frame relay switching of data, as compared to circuit switching for voice, and the ongoing conversion of traditional dial tone to Web tone.

Regardless of the technical aspects of this information revolution, its most prominent impact has been on the structure of organizational command, control, and communication systems. Fewer levels of management now oversee a broader span of responsibility and control over an increasingly international scope of activities. Global plant scheduling, global procurement systems, finished-goods inventories, and interunit logistics can now be directed and monitored from a fewer number of better informed global operating bases. Customer demand and distributor inventories can be routinely inputted into production schedules, and just-in-time inventory procurements more easily arranged. Additionally, current data on all these factors is immediately available for timely review at all levels of management.

Technology has also significantly impacted global transportation systems. Highly segmented, general-purpose domestic and international transport vehicles have now been replaced with seamless, intermodal, and highly specialized transport systems. Supertankers, high-speed container ships, and roll-on–roll-off ships have replaced previous general-purpose freighters. Wide-body, big-belly cargo, and passenger jet aircraft have vastly increased the capacity and reduced the cost of air shipment. Unit trains and piggyback railway cars, as well as many other highly specialized railcars, now outnumber conventional cars. Larger truck fleets, with bigger individual and/or tandem units, now travel significantly improved intercontinental highway systems, with fewer border controls.

Binding together these various transportation modes is an advanced intermodal transport infrastructure of redesigned and re-equipped seaports, enhanced cargo-handling airports, and better loading and unloading facilities for land-based transport systems. Today, it is not unusual for parts to leave a supplier's plant in a container, to be trucked to a seaport, travel an ocean, ride a train, be carried to their final destination on a truck, and ultimately be delivered directly to an assembly line on which those parts will be used, all in the same container in which the parts only recently left the foreign supplier's production line.

An important ingredient in the improvement of transport systems is also information-technology related. Through the use of bar-coded package and container labels, hand-held as well as fixed bar-code readers, and location-tracking computer systems, shippers can now enjoy the ability of immediately determining the current location of any of their shipments, and an accurate schedule for anticipated delivery. These improved shipment information systems have been one of the principal ingredients in the success and rapid growth of the package courier carriers, such as United Parcel Service (UPS) and Federal Express (FedEx).

Much as technology has benefited the emergence of the globalization paradigm, it has also substantially benefited from the very paradigm it helped create. Accelerated technological change has dramatically shortened product life cycles. The high cost of research and development coupled to these shortened product life cycles has mandated access to more and larger global markets in order to efficiently amortize the high cost of innovation. Quite frequently, the amortization of these costs represents a larger portion of total product and/or service cost than does the unit production cost. Selling more units during a product's active life is the obvious best solution, and accessing more and larger international markets is an effective methodology to increase such unit sales.

An early disincentive to the broad international sale of highly innovative products was the lack of uniform intellectual property right (IPR) protection in the world, ranging from developing nations, which in the past neither believed in nor recognized IPRs to other nations, which offered reduced levels of protection for shorter periods of time at greater levels of expense. Fortunately, in the grand bargain of the GATT Uruguay Round, improved IPR protection was one of the benefits achieved by the developed nations in exchange for more liberal and more secure access to developed-world markets by the nations of the developing world. This achievement is more fully described in Chapter 3, which deals with technology and innovation. By better protecting an innovator's rights to his own inventions, the incentive to engage in the high-risk process of research and development has not only been sustained but significantly enhanced.

Globalization has also gained from the efforts expended in the eight negotiating rounds of GATT, which were aimed at reducing import tariffs and other trade barriers between nations. The first seven rounds of the GATT concentrated primarily on reductions in import duties on merchandise trade between the advanced economies. However, in the seven-year-long GATT Uruguay Round, the agenda was broadened to include other subjects, such as trade in agriculture and services, allowable safeguard measures, phase-out of the Multifibre Agreement (MFA) quotas in textiles and clothing, government subsidies, the protection of IPRs, and trade-related investment measures. The substantial successes of these multiple GATT negotiating rounds, and the Uruguay Round in particular, have vastly expanded the global opportunities for trade and investment.

At the same time, the deregulation of domestic markets and the privatization of domestic-infrastructure service providers have significantly increased the level of domestic competition and the access of foreign competitors to market sectors formerly reserved for domestic enterprises. Of particular benefit has been the privatization of domestic-infrastructure service providers, which has permitted foreign direct investment in these entities. Once competition is introduced into these market sectors, the nationality of potential suppliers pales in importance to which vendor can provide the most conspicuous customer solution. In the GATT Tokyo Round, much negotiating capital was expended to establish a plurilateral agreement on government procurement. The privatization of state enterprises, in which foreign direct investment is welcomed, should be a better and more lasting solution to this issue than the Tokyo Round Agreement. Probably the best example of this phenomenon is in the global telecommunications carrier industry. The moment government ownership disappeared, so did the preference for indigenous procurement. Today, the most conspicuous customer solution is what sells in the world of telecommunications. Equally good examples exist in airlines, electrical utilities, water companies, and so forth. Potential deregulation that the United States might gainfully explore is the repeal of the current Jones Act, which discriminates in favor of U.S. maritime carriers in intra-U.S. shipping trade.

GLOBALIZATION MILESTONES

One way to measure the impact of globalization on the world economy is to compare the relative growth rate of aggregated world output to the growth rates of international trade and FDI. As noted in Chapter I, between 1991 and 2000, world output was expected to rise 3.2% annually, international trade 6.2% annually, and FDI about twice as fast as international trade. Thus, these two measures of globalization are growing considerably faster than world output.

As also noted in Chapter I, the IMF expects the annual exports of goods and services to rise to about U.S.$7.5 trillion by the year 2000. The United Nations (U.N.) expects that the sales revenues of FDI foreign affiliates will then exceed the exports of goods and services by a factor of 1.7, indicating that these FDI sales revenues would reach U.S.$11.5 trillion by the end of the twentieth century. Also by that time, the number of global parent enterprises could be approaching 60,000, and the number of FDI affiliates could be approaching 500,000.

An interesting aspect of the relationship between trade and FDI is that today, one-third of all global trade is intracompany, that is, trade between parent and affiliate or between affiliate and affiliate. For U.S. enterprises, the percentages for intrafirm trade are considerably higher. As regards royalties and fee payments for technology and trade names, today nearly 70% of these are intra-company.

In the minds of most businessmen, the world has entered a new era of global integration and rapid technological advance. As noted earlier, a similar high level of integration was achieved in the early years of the twentieth century. However, at that time most of today's businessmen were yet to be born.

To many, today's paradigm means that there has been a break-ing of the mold economically. Businessmen feel that the old par-adigms are no longer very helpful nor very useful. In fact, the old paradigms probably still do apply, but their metrics have changed. Zero-defect quality is no longer a goal; it is the minimum require-ment for doing business. Product differentiation is still important, but 10% to 20% improvements no longer qualify as differentiation; 100% to 200% improvements are now common. Time-to-market is now even more important, but it is no longer measured in months and years, it's now measured in day and weeks.

The world is changing from local, national, and regional econ-omies and industries to truly global ones. The Dow Jones Indus-trial Average (DJIA) has become, in reality, an indicator of how the world economy is doing, even though it only measures the performance of thirty large U.S.-based enterprises. This anomaly is explained by referencing a 1995 summary, published by the United Nations, which lists by foreign assets the top one hundred transnational corporations in the world. Of these one hundred cor-porations, twenty-seven are U.S.-based. Also included in the U.N. summary is what the United Nations calls a transnationality in-dex, which is the combined average of each company's percentage of foreign assets, percentage of foreign sales revenues, and per-centage of foreign employment. Taken together, these twenty-seven U.S.-based transnationals have a transnationality index of just less than 48%, with an index range of from 25% to almost 70%. Thus, the DJIA is today also a good indicator of just how the world economy is doing.

The new global paradigm places ideas and innovations, not things, at the center of trade and commercial success. In an idea-based economy, there is an obvious relationship between the size of the addressable market and the value of the innovative idea.

Globalization has aided U.S. industry to finance necessary investment at more affordable rates than would have been possible through financing investment solely from the low savings of U.S. citizens. Fortunately, there have been massive amounts of money on a global basis looking for investment opportunities. Jack Welch, the dynamic and extraordinarily successful chief executive officer (CEO) of the General Electric Co., expressed the challenges of this new global paradigm well when he said, "The real task of management is to allocate capital, both financial and human, and to transfer best practice."

GLOBAL COMPETITIVENESS

As postulated at the beginning of this book, the enlightened confluence of technology, innovation, trade, and investment is the route to superior levels of global competitiveness and "conspicuous customer solutions." What defines these superior levels, and do they apply to nations as well as to individual enterprises?

Nations do compete, but ordinarily not on the same basis as enterprises which compete for specific contracts or the global market share of specific products. Competition between nations more often deals in relative standards of living, the quality of life enjoyed by its citizens, the richness of national culture, financial strength and, unfortunately, often in military assets. But in the final analysis, the true measure of a nation's global competitiveness is the degree to which its institutions, laws, and regulations encourage the development of a national economic environment favorable to the nurturing of resources critical to the global competitiveness of its private sector.

The World Economic Forum (WEF) in Davos, Switzerland, measures a nation's global competitiveness based on a compendium of factors. How open is the nation's economy to international trade and finance? What is the role of government regulation? What is the degree of financial markets' development? What is the quality of national infrastructure, technology and innovation, business management, and human-resources development? What is the level of labor flexibility, and last, but certainly not least, what is the quality of the nation's judicial and political institutions?

Writing in a 1990 edition of the *Harvard Business Review*, Prof. Michael Porter expressed the view that "national prosperity is created, not inherited. It does not grow out of a country's natural endowments, its labor pool, its interest rates nor its currency values, as classical economics insists. A nation's competitiveness depends on the capacity of its individuals and enterprises to innovate and upgrade. . . . A nation does not inherit but instead creates the most important factors of production, such as skilled human resources or a scientific base. The stock of factors a nation enjoys at a particular time is less important than the rate and efficiency at which it creates, upgrades, and deploys them."

In seeking to describe the characteristics of a globally competitive enterprise, these would seem to be appropriate elements: a consistent source of innovative products and/or services and new ideas; a customer focus on the creation of customized solutions and highest-value products and/or services; and a least-total-cost producer offering unfailing zero-defect quality at least-elapsed time to market. And motivating this kind of performance must be a skilled management that believes in change, sets challenging corporate goals, and fully develops the capabilities inherent in its human-resource base.

Said another way, a globally competitive enterprise is one that consistently provides conspicuous customer solutions. In today's competitive environment, the big do not outperform the small, the swift outperform the slow. In the new paradigm of competition, least-elapsed time means that interval between identification of the customer need and the satisfaction of that need. This interval is much more inclusive than just the interval of the manufacturing function.

There are those who say that the above characteristics are desirable goals but that no enterprise can be expected to meet these rigorous standards on a consistent basis. Those who believe this are living in a past era or participating in market sectors not exposed to rigorous global competition. Not only are there many global enterprises active in today's market environment which consistently fulfill these objectives, but in many sectors these goals are really the minimum requirements for market entrance and participation. Those that would consistently fail these standards in markets exposed to intense global competition are best advised to not apply.

The disciplines of commerce are in transition. The former regional focus is moving toward a global focus. Worldwide agreements are replacing geographically suited contracts. Regional business terms are giving way to international compliances, such as ISO standards. Commodity purchases are being upgraded to technology investments, and the purchase of parts is being extended to the management of the parts-supply process. A total cost/value focus is replacing the former piece/part focus, and a conspicuous customer solution focus has replaced product focus.

Through the integration of the above methodologies, enterprises now seek to harvest the ultimate advantage inherent in the globalization paradigm, so succinctly defined by Peter Buckley in 1989 as the ability "to convert global inputs into outputs for global markets as efficiently and profitably as possible."

Chapter 3

Technology and Innovation

As noted in the foreword of this book, the United States has long been the acknowledged leader in the generation of technology, and, through the process of innovation, the timely conversion of this new knowledge into conspicuous customer solutions. CED's 1998 policy statement, entitled "America's Basic Research-Prosperity through Discovery" quotes a former secretary-general of the Royal Swedish Academy of Sciences, responding to the question of why so many Nobel prizes have been won by Americans, as saying "no other country has invested as much in research over the years as the United States. It's as simple as that." Reminiscent of the secretary-general's comment was a favorite saying of my late grandfather, to the effect that if the ability to play good golf was dependent upon the expense of your golf clubs, the rich people would be the world's best golfers. Obviously, there must be other relevant factors to generating new knowledge and to playing good golf.

During the last half of the twentieth century, the United States has been the major producer of new knowledge and its innovative application. In the early years following the end of World War II, the United States was probably the source for at least 75% to 80% of the world's new knowledge. In the 1990s, the U.S. share has probably fallen to around 50%, which in no way indicates that the

U.S. production of new knowledge has decreased. As other economies have recovered from the war, their production of new
knowledge has significantly increased, as has their ability to innovate new solutions based on technology purchased or licensed
from the United States.

By the late 1990s, nearly U.S.$200 billion was being spent an-
nually in the United States on research and development. Of this,
nearly one-third was funded by the U.S. government. Of the U.S.
$25 billion spent on basic research, the government's share was
almost 70%, primarily in the areas of defense, space, and health,
in which the U.S. federal government has a mandated role. For
many years, the U.S. government's major participation in the
funding of domestic R&D has significantly concerned many U.S.
trading partners, who have looked upon such funding as a subsidy to U.S. exporters, particularly in the rapidly growing markets
for high-tech products. Many foreign government research-and-
development subsidies, such as those to the producers of Airbus,
have been justified as a legitimate response to the U.S. research-
and-development subsidies to high-tech industries.

In view of this global concern, it is quite interesting that in the
GATT Uruguay Round the panel on export subsidies decided that
the promotion of research and development was a legitimate government activity. There was obviously much pragmatism in this
call. It would have been difficult for the new WTO to defend the
position that governments should not be allowed to participate in
the funding of R&D addressed to the solution of important global
problems. Additionally, except for products such as military hard-
ware, nuclear power generation, and health care, the R&D being
funded by the U.S. government has become considerably less relevant to commercial application, and then only many years down-
stream.

As these debates on R&D subsidies raged, numerous attempts
were made to compartmentalize the process of invention into areas such as basic research, applied research, development, and
implementation. As invention moves from basic research to implementation, the inherent risk in each step is significantly lessened, but the funds consumed are significantly increased.

For purposes of this book, I do not believe such exercises in
compartmentalizing the lineage of new ideas is all that enlight-
ening. Instead, this discussion will concentrate on the rather all-

inclusive terms of technology and innovation and not be too concerned with how the process of new-idea generation gets assigned to either of these two broad categories.

TECHNOLOGY

Fundamentally, technology is the knowledge of how to perform useful tasks and how to create useful products. Much technology is in the public domain and is not protected by an intellectual property right. In many cases, this fundamental level of technology is what is sought by developing nations. But even though this technology does not have a specific owner, the developing nations still need an avenue by which to garner it. In many cases, it is the process of foreign direct investment that best accomplishes this mission, even though it is certainly not the only avenue available.

Technology relates to knowledge, often new knowledge, and frequently new scientific knowledge. In essence, technology constitutes an inventory of potential value, but only if it is exploited on a timely basis. A good way to think about technology is as the intellectual headwaters of innovation, a raw material for creating change.

Innovation, on the other hand, is the conversion of technology into conspicuous customer solutions, which can be either product, process, or service. Innovation is the process by which technological value is realized. High-value innovation is dependent upon the enlightened definition of a significant problem and, on a timely basis, the coupling to that problem of an adroit technological solution.

Unfortunately, in recent years, the impression has been created that innovation is limited solely to high-tech products and high-tech industries. Nothing could be further from the truth, as innovation is applicable to any product, any process, or any service in any field of endeavor. Innovation is applicable to marketing, cost leadership, zero-defect quality, customer service and elapsed time. You can innovate in finance, although at times it does seem as if there has been a bit too much innovation in finance.

Good examples of non-high-tech innovations are Domino's Pizza and FedEx. Tom Monahan did not invent the pizza, but he revolutionized the pizza industry with guaranteed fast delivery.

He improved the business economics by eliminating the need for numerous expensive restaurants, located on high-priced real estate, and replaced these with a few strategically located kitchens within large metropolitan areas, from which teenagers in their own cars could race to deliver the pizza, oven-hot, within the guaranteed time limit. His innovation was not the product, but innovative customer satisfaction and a more economic investment base.

Although Fred Smith's idea to create FedEx used high-tech equipment such as jet aircraft, automated sorting systems, and a state-of-the-art package tracking and communication system, his basic idea was very much simpler and at the same time extraordinarily innovative. He found that market in which the overnight delivery of a document was worth U.S.$20 as compared to the U.S.$0.20 cost for delivery by the U.S. Postal Service at an undetermined future date. He also found a companion market for the expedited delivery of packages. Although many had postulated the cost advantage, both real and physiological, of expedited package delivery, Fred Smith's innovation proved it. It could be said that both Tom Monahan's and Fred Smith's innovations were in marketing and the reduction of elapsed time-to-market.

INNOVATION

Despite the broad consensus on the critical importance of innovation, the available contemporary wisdom on how to innovate is much less instructive. You can obviously educate people in technology, but can you teach curiosity, an important characteristic common to innovators? Why is one enterprise innovative when others are not?

Intelligent people are obviously a major source of innovation. Combining intelligent people into groups of sufficient size to achieve the cross-stimulation of new ideas is also a productive practice. But the common denominator of eliciting quantum innovation is pressure. To generate high levels of innovation you must ask for it.

The intensified level of competition achieved through the new paradigm of globalization is a major factor in increased global innovation. Even though this global competition puts increased pressure on almost every global enterprise, not all enterprises have enjoyed equivalent success rates in innovation. So what has been the differentiating factor?

The late Walter O'Malley, longtime owner and president of the Brooklyn and later Los Angeles Dodgers, had a favorite expression that speaks directly to this point. According to Walter, "Fish smells from the head"—a graphic way to describe the importance of top management's involvement in the process of nurturing innovation. Unless the CEO embraces the critical importance of innovation, neither will others in the enterprise, including the potential innovators.

This is not to say that the CEO must have the technical knowledge, vision, or skills of a Bill Gates or Gordon Moore to be capable of stimulating innovation. But the CEO must be the corporate cheerleader for innovation, and he must set enterprise goals that require innovation in order to be satisfied. He must also recognize that although pressure must be applied to elicit innovation, it will not appear broadly across an entire enterprise, but will appear in pockets.

Furthermore, it is important to realize that these valuable corporate assets, which innovators represent, must be protected from those who feel threatened by them. Innovation is frequently not viewed positively by the corporate bureaucracy, which often fears change and frequently moves effectively to contain it. This is, of course, one of the major issues faced by an innovative enterprise, which, because of its past ingenuity, experiences rapid growth. With that growth comes an enlarged corporate bureaucracy skilled at resisting change, with which it is uncomfortable. Thus, just the creation of corporate goals that require change is not enough.

The potential agents of innovative change must be identified, encouraged, equipped with the necessary resources, and protected. The normal corporate reward system, such as new titles or broadened spans of control, are not necessarily the style of recognition sought by highly imaginative innovators. Top management interest in these innovators, assistance on their projects by designated top management project champions, and broad enterprise recognition of innovative accomplishments are much more effective. Regularly scheduled technology days, when specific innovators enjoy the opportunity of displaying their new works to a broad spectrum of corporate management, can be a very effective methodology and often exert a strong influence on future business plans.

Innovation is a demanding task, and not all days are good days. Innovators often experience multiple failures on the road to a successful solution. A high-innovation success rate may often be an indication that the goals for innovation have not been set sufficiently high. Because of this tedium factor, frequent inquiries and expressions of interest in an innovator's projects can often apply the pressure necessary for sustained innovation. But just as in merchandising, where the three most important factors are location, location and location, in innovation they are pressure, pressure, and pressure. Those enterprises that effectively apply internal pressure to complement the external pressure of intense global competition are those most likely to foster successful innovation.

INNOVATORS

It is impossible to define a typical innovator or what causes innovators to innovate. Innovators come in all varieties and are motivated by a wide spectrum of influences. There are, however, certain characteristics that most innovators share. First, and most important, innovators believe change will occur and that it will be swift. They believe that change is predictable and that it is capable of objective analysis.

Innovators concentrate on being in the right technologies, at the right time, and with the right people. They believe that innovation is more important than being ever more efficient in current activities. This is not to say that innovators fail to recognize the importance of being ever more efficient in current activities. They well recognize the importance of continuous incremental improvement in each and every facet of a business. What they do believe is that such incremental improvement alone is not enough; it must be paralleled with activities seeking bold new initiatives.

Innovators regard product quality, product differentiation, and interval-to-market as the vital metrics of the innovation process. In this regard, their beliefs are not accidental. They are based on understanding the dynamics of competition, understanding the

Figure 3.1
The S-Curve

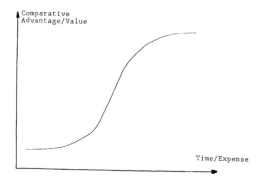

attacker's advantage, understanding physical limits, and understanding the theory of the S-curve.

THE S-CURVE

Figure 3.1 portrays an S-curve, which takes its name from the
fact that the curve looks as if someone grasped the two ends of
the letter *S* and pulled them in opposite directions. What an S-
curve describes is the changing relationships, during the pursuit
of a new idea or new technology, between the comparative advantage and/or value created and the time and/or expense incurred in creating that value.

The pursuit of a new idea or technology begins in the lower,
left-hand quadrant. Initially, units of time and expense are invested, but little or no additional value or comparative advantage
is created. However, after a time these inputs begin to combine
and value begins to be created. Soon value begins to rise at a rate
faster than time and expense are being added. This is the "golden
era" of innovation, and to some, less than fully experienced innovators, there is the hope that this era will continue forever.
However, as the experienced innovator so well understands, ultimately new ideas and/or new technologies begin to approach
their physical limits. As these ideas and/or technologies begin to
approach their physical limits, the creation of additional value begins to decline in relation to the addition of time and expense.
When the physical limit is finally reached, no amount of addi-

Figure 3.2
Multiple S-Curves

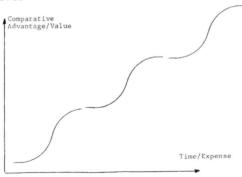

tional time or money will create additional value. It is in this period of a declining ratio between value created and resources expended that the experienced innovator begins to lose interest in the present idea and/or technology and initiates a search for a successor solution.

It is at this point when most competitors are still trying to squeeze the last ounce of advantage from a mature technology and more innovative competitors are already seeking the next solution, that market leaders are determined, that market share is gained or lost, and that certain products and companies begin to suffer reductions in sales revenues and product margins.

Figure 3.2 displays the activity of the most facile innovators, who when they begin to experience a declining ratio of value-added to expense and time, immediately move on to the identification and exploration of a new, more promising technology or idea. Such a move obviously can be most wisely made when the innovator is well aware of the absolute physical limits of his current approach. There are many examples of such innovative technology hopping: the move from electro-mechanical switches to vacuum tubes to semiconductors, the evolution of aircraft engines from piston to turboprop to pure jet engines, and the migration from phonograph records to magnetic tapes and compact discs. A good domestic example is the innovation of disposable diapers, which replaced "dydee" washes, which replaced home-washed diapers.

A good question raised by the S-curve is the appropriate point at which to introduce new ideas to market. Not every S-curve has the same shape. For some, the time axis is longer than on others, and on some the value gain is sharper. In general, the market introduction point should be the point at which maximum value has been created in the least elapsed time. Depending upon the shape of the specific S-curve, that point is probably reached in the portion of the curve at which the curve is just beginning to bend and create less new value for additional increments of time and expense.

This point could vary based on existing market conditions, on relative market shares of current competitors, and on a seat-of-the pants feel of just what must be done to ensure being first-to-market. There is the old adage that the first competitor cleans up, the second does okay, the third breaks even, and the fourth loses his shirt. Dominant competitors also enjoy the luxury of announcing a new idea in the early phases of value creation, with a promise to deliver greater value at some future date certain. This often has the effect of freezing the current market and causing current customers to await the arrival of the differentiated product, rather than continuing to purchase the current models. The validity of this market approach is highly dependent on the reputation of the supplier to deliver as promised. It also requires the ability to accurately predict the future evolution of the specific S-curve and the point at which the idea's physical limits will come into play.

THE GLOBAL IPR ISSUE

In today's world of low-cost and rapid communication and information systems, it is obvious that technology and innovation travel quickly, globally, and in a fashion that can significantly shrink an innovator's window of comparative advantage. Since technology and innovation are so important to both global competitiveness and the growth of nations as well as individual enterprises, it is most important that incentives be offered to encourage innovation. And there is no stronger incentive to undertake the high-cost, high-risk process of research and development than to allow the innovator the exclusive right of ownership of his or her innovation for a reasonable period of time, better known as an intellectual property right. Without such rights, who would want to undertake this risky, expensive process of innovation?

For many years, the United States has had a sound intellectual property right protection system, which includes patents, copyrights, and trademark and styling protections. In addition, some innovators choose to protect their inventions through the process of trade secret, particularly when the disclosure of the invention required to achieve formal protection would reveal sufficient information to enable someone to develop a solution that avoids the protection granted. This is particularly true in process patents, and is the reason for which the Coca-Cola syrup process has remained a trade secret. Obviously, chemists can determine what is in the syrup, but they can't determine how the mixture is created.

The instruments of intellectual property right protection can be used in several ways. Most obvious is their ability to reserve to the right's owner the exclusive right to make, use, and/or sell the innovation. The right can also be licensed to others in situations when the owner may not wish to commit the capital to exploit the innovation, the innovation falls beyond the owner's sphere of business interest, or the owner feels someone can better develop a specific market with which the owner may lack familiarity.

Today, a major use of such instruments is in trading them for a right held by someone else. Complex innovations often encounter prior art in some portion of the new solution that is already IPR protected. The owner of that prior art may not be interested in licensing it, but could very well be interested in trading his right for the right to some other intellectual property. Thus, broad portfolios of tradable IPR instruments, covering multiple areas of innovation, are desirable insurance against being boxed into a solution that is blocked by someone else already owning a right to some portion of that solution.

Many other developed nations offer intellectual property right protection, but often at lower levels, for shorter periods of time, and at higher costs than in the United States. Historically, the developing nations have represented the major problem in the protection of intellectual property. In the past, it was the view of the developing world that technology is part of the common heritage of mankind and, therefore, like any other form of knowledge that lies in the public domain, it should not command a price.

The developing world also believed that the costs of private research and development should be largely recovered from the advanced economies and that the developing nations' more limited capacity to pay entitles them to a lower price than that charged to other purchasers. Additionally, the developing nations regarded technology as overpriced because it was transferred under monopolistic conditions, and because the expenses incurred in developing it had already been sunk. This last view is particularly flawed in that the revenues from current technology are necessary to finance a continuing flow of new technology. The correct view of this circumstance is that the development costs of current technology were paid for by the previous generation of technology, and the revenues of current technology are intended to pay for succeeding technology.

U.S. innovators are entitled to file for intellectual property right protection in every country in which they believe their invention has a sales potential that would justify the time and expense of such filing. However, they are not guaranteed the U.S. level of protection in those countries. They are entitled only to the level of protection offered in each country, assuming their invention can qualify domestically for the protection sought. Because of the many developing countries in which little or no IPR was offered, the globalization paradigm potentially jeopardized the existing IPR systems of the developed world.

Inadequate global IPR protection can create significant international trade distortions, and it has the potential to displace sales of legitimate goods in the innovator's domestic market. It has the potential to reduce exports, foreign sales, and royalty fees in foreign markets that abuse IPRs, and it can shrink the market for legitimate goods in third-country markets through the export of counterfeit goods from nations that abuse IPRs. To protect the IPR rights of U.S. innovators in the U.S. market, in the Tariff Act of 1930, the U.S. government included Section 337, which specifically authorized the U.S. International Trade Commission to exclude imports into the United States that violate U.S. intellectual property right statutes.

In spite of this protection afforded IPRs in the U.S. domestic market, in the early 1990s the International Trade Commission (ITC) estimated that foreign IPR infringements could still annually cost U.S. exporters up to U.S.$50 billion in lost foreign sales. In negotiations between developed and developing nations regarding this issue, the developing nations favored a solution of national treatment. However, if, as in most cases, these developing nations offered little or no IPR protection to their domestic innovators, it was not of much value to extend such nonprotection to foreign innovators. Thus, when the GATT Uruguay Round was initiated in 1986, IPR protection was one of the principal developed-world objectives in the round's grand bargain with the developing world. From the developed-world's point of view, the accomplishments pertaining to IPRs, achieved in the Uruguay Round, were among the round's most important successes. The GATT agreement on IPRs was also facilitated by a changing view as regards IPRs in the developing world. As individual nations in the developing world began to industrialize and recognize the growing importance of innovation and new ideas as contrasted to things, their views of IPRs began to soften. However, to this day, paying for technology is still not a popular requirement in many developing nations.

IPRs AND THE GATT URUGUAY ROUND

The IPR achievements within the GATT Uruguay Round, better known as trade-related intellectual property rights (TRIPS), were truly remarkable. The agreement established a national treatment commitment under which foreign nationals must be given IPR protection equivalent to that accorded a country's nationals. A most favored nation (MFN) clause was also included, as were certain basic minimum protections to be accorded to intellectual property.

With respect to copyright, all GATT signatories are now required to comply with the provisions of the Berne Convention of 1971. The TRIPS adopts Berne's minimum period of life of the author plus fifty years for protection of works, except for the obligation with respect to moral rights (i.e., the right to object to damaging revisions or displays of creations). Computer software and databases are specifically protected as literary works under the Berne Convention. Thus, computer programs are protected for not less than the life of the author plus fifty years.

TRIPS also expands and clarifies the Berne's disciplines by providing that the term of protection should not be inferior to fifty years in cases in which the term is calculated on a basis other than the natural life of a person. This deals with the situation of works that belong to corporations, where the term "is no less than fifty years from the end of the calendar year of authorized publication, or failing such authorized publication within fifty years from the making of the work." As regards anonymous (or pseudonymous works), the term of protection in Berne (and thus in TRIPS) is fifty years after the work has been lawfully made available to the public.

Finally, in 1998, the 105th Congress passed an extension of the copyright term that brings U.S. practice in line with that of the European Union (EU). The new law extends the duration of a copyright in a work created on or after January 1, 1978, to the life of the author plus seventy years after the author's death. It also extends the duration of copyright in anonymous or pseudonymous or works-for-hire on or after January 1, 1978, to ninety-five years from the year of first publication, or 120 years from the year of creation, whichever expires first. These new copyright terms of protection are not required by the Uruguay Round agreement, but are U.S. law.

The agreement further defines what types of signs must be eligible for protection as a trademark or service mark and what the minimum rights conferred on their owners must be. Industrial designs or styling are now protected for a period of ten years. Owners have the right to prevent the manufacture, sale, or importation of articles bearing or embodying a design that is a copy.

As regards patents, there is a general obligation to comply with the Paris Convention of 1883, as revised in Stockholm in 1967. In addition, the agreement mandates twenty-year patent protection from the date of filing for all inventions, whether process or product, in all but a very few fields of endeavor.

The agreement establishes a council for trade-related aspects of intellectual property rights to monitor the operation of the agreement and compliance with it. Developed nations were given until January 1, 1996, to comply. Developing nations and countries in transition were given until January 1, 2000, and the least developed nations were given until January 1, 2006, to comply.

TECHNOLOGY, INNOVATION, AND GLOBALIZATION

As this chapter has emphasized, technology, innovation, and globalization are closely intertwined. Each impacts the other and is impacted by the other. The growth of each has been aided and benefited by the other. Fortunately, through measures such as the GATT Uruguay Round agreement on IPRs, the ability of each to continue to support the other has been enhanced. In the following chapters we will explore how international trade and foreign direct investment, the complementary agents of globalization, support the process of expanded global commercial intercourse and the creation of conspicuous customer solutions.

Chapter 4

International Trade

The previous chapter reviewed the importance of new ideas, technology, and innovation to the creation of conspicuous customer solutions. It also emphasized the close relationship between technology, the asset, and innovation, and the process for realizing value from this asset. Additionally, the impact of globalization on new ideas was explored, particularly as regards the intellectual property rights to these new ideas.

In this chapter, the intent is to demonstrate the additional link between new ideas and the complementary business tactics of international trade and foreign direct investment, utilized to achieve the optimal global deployment of an enterprise's resources in its quest to convert global inputs into outputs for global markets just as efficiently and effectively as possible. Specifically, this chapter will explore the subject of international trade and its contributions to conspicuous customer solutions.

As noted earlier, international trade is expected to grow to about U.S.$7.5 trillion annually by the end of the twentieth century. Additionally, in recent years, international trade has been growing at a rate twice as fast as world output. Politicians, labor leaders, and too many businessmen praise increased trade solely because it means more exports, which means more jobs, usually "good" jobs. Likewise, these same groups frequently urge consumers to favor domestically made goods over imported goods.

Economists, on the other hand, believe the real benefits of trade lie in importing rather than exporting. James Mill, who was one of the first trade theorists, explained this phenomenon in 1821: "The benefit which is derived from exchanging one commodity for another arises, in all cases, from the commodity received, not the commodity given." From James Mill's comments we can postulate that the optimal role of international trade in the world economy is, therefore, to provide, through the maintenance of open world markets, broad-based improvements in global economic welfare. The capability of trade to contribute to economic welfare derives from the three dynamics inherent in the process of international exchange: specialization, competition, and economic adjustment.

Specific nations and specific enterprises enjoy comparative advantage in the production of certain products and services. By concentrating on the production of these products and services for both domestic consumption and export, these nations and enterprises can achieve economies of scale and the optimal utilization of national and enterprise resources. By importing products and services in which other nations or enterprises enjoy comparative advantage and economies of scale, these nations and enterprises can enhance their economic efficiencies. Such practices reinforce the concept of converting global inputs into outputs for global markets just as efficiently and effectively as possible.

Increased global competition creates greater pressure for excellence, new ideas and innovation, and new investment. It also creates greater pressure on costs and prices. Increased global competition augments domestic competition by creating pressure for improved performance.

And what is the importance of economic adjustment? The global market is dynamic, not static. Products and processes change; nations and enterprises evolve in different ways and at different rates. Thus, specific comparative advantage can shift both in degree and location. Economic adjustment is necessary to ensure that nations and enterprises react to shifting comparative advantage, and that resources be moved continuously toward centers of comparative advantage so that they continue to be optimally employed.

WORLD TRADE HISTORY

The international exchange of goods and services has its roots in the very beginning of the world economy. Long before the Industrial Revolution, Europeans traded for spices and other exotic products with partners as far away as Asia. In the fifteenth and sixteenth centuries, the British, Dutch, Spanish, and Portuguese developed the concept of mercantilism, as previously mentioned, based on large ocean fleets and large colonial empires. World trade really began to flourish in the eighteenth and nineteenth centuries as transport costs fell significantly and Britain initiated more liberal trading policies. During the later portions of the nineteenth century, other world nations followed the British lead. Trade continued to grow, and the world continued to integrate economically up to 1914.

For the next thirty years, the world economy suffered severe dislocations, which set back further gains in trade and economic integration. These dislocations included World War I (1914–18), the Great Depression (1929–38), and World War II (1939–45). The tensions spawned by the Great Depression are generally considered one of the major triggers for World War II, and inappropriate global trade policies are believed to have prolonged the Great Depression and accentuated its impact.

Among the most damaging trade polices enacted during the Great Depression was U.S. legislation introduced in 1930 by Sen. Reed Smoot, Republican of Utah, and Representative Willis Hawley, Republican of Oregon. Known as the Smoot-Hawley Trade Act of 1930, this bill significantly raised all U.S. agricultural and industrial import tariffs. The result was predictable. U.S. trading partners responded with similar measures, world trade declined, global economic integration atrophied, the Great Depression was prolonged, and contributing tensions to the initiation of World War II were intensified.

The Smoot-Hawley Act was an important economic lesson for the United States. But as fewer and fewer people in today's society remember the 1930s, much of the Smoot-Hawley lesson is being lost. To those who currently say that the United States can no longer afford liberal trade policies, the correct response must be: We've been there, we've tried that, and we still have the tragic scars.

As World War II wound down, the British and American architects of the postwar economic era recognized the need to reintegrate the global economy and organized several new international institutions. Included in these were the International Monetary Fund (IMF) and the International Bank for Reconstruction and Development (IBR&D), for finance, and the International Trade Organization (ITO), for world trade. The IMF still exists, the IBR&D is now called the World Bank, but the ITO never came into existence because the U.S. Congress refused to ratify it. The ITO was, therefore, replaced in 1947 by the General Agreement on Tariffs and Trade, which essentially consisted of the commercial policy chapter lifted from the failed ITO.

It is important to recognize that the first thirty years of GATT were the most prosperous in the world's history prior to the 1990s, even though trade liberalization under GATT was largely confined to the developed or advanced economies and to nonagricultural merchandise trade.

The GATT Uruguay Round, which was its eighth negotiating round, and which was concluded in 1994, created the new World Trade Organization. The WTO, in effect, succeeded GATT, significantly strengthened world trade disciplines, and essentially gave the world an ITO equivalent, which the U.S. Congress declined to ratify in 1946. Thus, as the world prepared to exit the twentieth century, world trade and global economic integration had returned to those levels achieved in the early years of the century. And thanks to the pioneering activities of GATT and other international institutions, the process of globalization had become more friendly to the further achievement of conspicuous customer solutions available to a broader segment of the world's population.

TRADE AND THE 1990s

The 1990s achievement of higher levels of world trade and global economic integration was aided by a number of important trade developments that occurred during the decade. Most important was the very successful conclusion in April 1994 of the Uruguay Round of the GATT, and the creation of the new World

Trade Organization (WTO) in Geneva, Switzerland, on January 1, 1995.

Additionally, the European Union achieved a single internal market on January 1, 1993, and in November 1993, the U.S./Canada Free Trade Agreement was expanded into the North American Free Trade Agreement to initially include Mexico and potentially other Latin American nations. And finally, talks were initiated to create an Asia/Pacific Economic Cooperation forum (APEC), a Western Hemisphere Free Trade Area (WHFTA) or a Free Trade Area for the Americas (FTAA), and, last but not least, talks took place as regards the creation of a North America/EU trans-Atlantic free-trade area. There are divergent views as to whether such regional trading arrangements constitute building blocks or stumbling blocks to a truly open world-trade system. However, it is hard to argue with the economic, social, and political successes of the EU. Nevertheless, the proposed EU/NAFTA agreement needs careful thought as regards the position in which it places Japan, the world's second largest economy.

Underlying these specific trade developments during the 1990s have been even more important long-term trade trends. The decade has witnessed the accelerating globalization of markets not only for goods but also for services, capital, technology, and investment. There has been an extraordinary worldwide shift in both ideology and policy toward market-driven rather than government-directed solutions to national economic problems in the nations of the developing world and the countries in transition. Regional economic crises late in the decade did create certain, but hopefully temporary, backsliding from some principles of market economics in a number of the emerging nations. And last, but certainly not least, there has been a blurring of the traditional North/South divide between the have and have-not nations, as many developing nations have diversified away from concentration on primary commodities and become competitive exporters of manufactured goods.

During this decade, concerns in the United States relating to trade have intensified because of the endemic U.S. current-account deficits, which are estimated to total over U.S.$1.3 trillion during the decade. So just as the U.S. federal budget deficit had been the primary U.S. domestic economic issue, the U.S. current account deficit remains its primary international economic issue.

Surprisingly, U.S. trade policy probably has played quite a small role in the decade's current-account deficits. Much more important have been the U.S. high-consumption/low-savings lifestyle, and the U.S. prosperity during the decade juxtaposed the less vibrant economies in the EU and Japan, which accentuated the U.S. demand for imports and lessened demand for U.S. exports.

TRADE POLICY

In a perfect world economy, there would be no need for government trade policies, nor for government trade interventions. The market system (prices) would guide optimal resource allocation, global market-based competition would prevail, the factors of production would be mobile between nations, and economic adjustment would be smooth and swift. But, alas, the world economy is not perfect and, thus, there is a need for trade policy on the part of individual nation states.

Trade policy is best defined as government measures that have industry-specific application. Whether or not applied at national borders, these measures have the ability to significantly affect international commerce in both goods and services, international investment, and the protection of intellectual property rights. Trade measures typically applied at borders are import quotas, tariffs, border taxes such as value-added taxes (VATs), restrictive technical, health, and safety standards, plus nontransparent measures of administrative guidance. Such measures can be said to be the new idea or innovative element in trade policy. Trade measures not applied at borders can include domestic subsidies to enterprises, such as to the producers of Airbus, domestically-oriented government agency or state enterprise purchasing policies, inadequate IPR protection, and local competitive practices such as cartels and the *keiretsu* groups in Japan, which grant procurement preference to other group members. Trade policy can dramatically influence the level of a nation's trade. A nation with high trade barriers invites similar high trade barriers from its trading partners. Thus, the volume of that nation's trade will be reduced. In that government, trade interventions are industry-specific; trade policy can obviously influence the composition of trade. But, can trade policy, as many believe, influence a nation's trade balance? Since most trade interventions are usually matched by offsetting interventions on the part of a nation's trading partners, the answer must be that trade policy has only a limited influence on a nation's trade balance. Therefore, under a regime of flexible currency-exchange rates, trade policy tends to affect primarily the level and composition of a nation's trade rather than the overall balance between its exports and imports.

And herein lies a most important factor in the contribution of international trade to the more facile creation of conspicuous customer solutions. Does national trade policy promote global economic welfare through the dynamics of specialization, competition, and economic adjustment, or does it blunt access to these dynamics through policies of domestic protectionism, which stifle the ability of enterprises to convert global inputs into outputs for global markets just as efficiently and effectively as possible? In the decade of the 1990s, the net influence of trade policy would seem to have been favorable to the further globalization of commerce.

If trade policy is perceived to have only a limited influence on a nation's trade balance, what factors can influence national trade balances? Without doubt, the first and foremost factor is the global competitiveness of that nation's private sector, which, in turn, is driven by capabilities such as innovation, costs, quality, speed of response, and customer focus. Government fiscal and monetary policy are important, as they influence domestic demand, and government currency-exchange-rate policy is important as it influences the global competitiveness of its domestic enterprises. Unfair trade practices by a nation's trading partners that distort market-based trade flows can also influence trade balances. Trade balances depend heavily on macroeconomic factors such as the balance between savings and investment, which in turn influences the global competitiveness of a nation's private sector.

BASIC TRADE PRINCIPLES

The post–World War II experience tells us that open world markets can best achieve broad-based gains in global economic welfare through specialization, intensified global competition, and

swifter adjustment to economic change. From this we can frame
the basic principles of a productive trade policy that best supports
the conversion of global inputs into outputs for global markets
just as efficiently and effectively as possible.

First, trade policy should facilitate the optimal application of a
nation's resources through efficient specialization. Second, trade
policy should promote global competition rather than impede it.
And third, trade policy should facilitate economic adjustment to
change rather than retard it.

Following the end of World War II, the United States has been
the principal world leader in espousing an open multilateral trading system, and in supporting first GATT and now the WTO.
However, beginning in the 1970s, under the increasing pressure
of endemic U.S. current-account deficits, the United States did begin to espouse a philosophy of "fair" rather than "free" trade and
the need for "a level playing field."

On some occasions, the United States will support an open multilateral trading system. At other times, however, it has used the
rhetoric of "fair trade" as the screen behind which to follow protectionist practices. Although the United States has not abandoned
its commitment to GATT (and now to its successor, the WTO), a
series of recent decisions has shifted U.S. trade policy away from
an almost exclusive reliance on multilateral procedures to a much
greater reliance on bargaining bilaterally with individual countries
or small groups of trading partners.

Over the years, U.S. trade laws have increasingly acquired protectionist overtones, and threats of protectionist remedies have all
but forced trading partners to accept other bilateral measures,
such as voluntary export restraints, to avoid harsher trade-remedy
alternatives. Typical of these was the Orderly Marketing Agreement (OMA), negotiated with Japan, which established a 1.65-
million annual quota for the importation of Japanese cars into the
United States.

To understand this protectionist shift in U.S. policy, it is necessary to understand the domestic politics of trade policy, which
are antithetical to the creation of conspicuous customer solutions.
The political debates as to when import competition should be
restrained are usually cast as "us—the domestic good guys" versus "them—the ugly foreign exporters," when the real issue is
"us—the domestic producers who get protected" versus "us—the

domestic consumers who must buy the protected products at higher prices." Under a policy of trade protection, national resources are no longer optimally employed, and a transfer of income occurs from the domestic consumers to the protected domestic producers.

Thus, the beneficiaries of trade protectionism are the narrow, concentrated groups that produce the protected goods, and the losers are other domestic producers, perhaps potential exporters, who use the protected goods and the domestic consumers. Since restrictions on trade lead to higher prices that domestic customers must pay domestic producers for the same product they could buy from foreign producers at a lower price, the best way to view protectionism is mainly in terms of who in the nation benefits from protectionism and who pays its cost, rather than in terms of national interests versus foreign interests.

Since the potential rewards from protectionism are so large, these concentrated producer groups, often aided by their labor unions, have great motivation to organize and lobby the U.S. Congress and executive branch for protection. The alarming size of political action committee (PAC) funds and "soft money" funds, which flowed into federal election campaigns in 1992 and 1996, were frequently sourced from groups seeking access to trade protection.

The users of potentially protected goods are diverse, they purchase many other goods than the ones for which protection is sought, they have less at risk, and thus less motivation to organize and lobby. Consumers are seldom able to create much counterpressure against protectionism. So where is the balance of political power on trade protectionism? The protectionism debate is never fought on a level playing field. The advantage is always to the advocates of protectionism, and the losers are frequently those who wish to convert global inputs into outputs for global markets just as efficiently and effectively as possible.

An antidote for trade protectionism's lack of a level playing field is the technique of multilateral trade negotiations as initially sponsored by GATT and now by the WTO. A multilateral negotiation mobilizes the support of global export industries that seek broad international market access for their conspicuous customer solutions, and that seek reductions in foreign-market barriers as the quid pro quo for reductions in domestic-market barriers. Thus, a multilateral negotiation makes it possible to supplement the weak consumer influence in combating protectionism by bringing to bear the pressure of the more dynamic export industries as a counterweight to domestic protection forces. As noted earlier, GATT has been the principal multilateral instrument through which the United States and the world have sought to improve the global trading system ever since its inception in 1947, and up to the time of its replacement by the WTO in 1995. In essence, GATT has been the post–World War II antidote to Smoot-Hawley type policies. When formed, GATT had twenty-three members, primarily the advanced economies. By the conclusion of the Uruguay Round, GATT membership had grown to over 120 nations.

THE GATT

In its thirty-eight-year history, the three principal functions of GATT were to sponsor multilateral negotiation rounds for the reduction of trade barriers, to establish and administer a set of rules for the conduct of international trade, and to provide facilities for the settlement of trade disputes.

GATT was a long and complicated document, but was based on comparatively few fundamentals, which included nondiscrimination in trade (imbedded in the most favored-nation principle), national treatment regarding internal taxation and regulation, the use of tariffs as opposed to quotas for the protection of domestic industry, the avoidance of unfair trade practices, and a willingness to enter into multilateral negotiations for the reciprocal reduction of trade barriers. The GATT Uruguay Round was the eighth and final GATT negotiating round.

The Uruguay Round was GATT's most ambitious undertaking, and it sought to do more than merely reduce tariffs on industrial products, the major goal of the previous seven GATT rounds. But even in this tariff reduction role, the Uruguay Round met with conspicuous success, as the resulting bound tariff rates are scheduled to fall, on average, almost 40%, compared to the original 33% reduction target.

In addition to the tariff reduction objective, the Uruguay Round sought new international rules for trade in services, trade in agriculture, and for the elimination of the Multifibre Agreement quotas on trade in textiles and clothing. It sought to harmonize technical, health, and safety standards, often used as trade barriers, and further sought to eliminate trade-distorting investment measures, GATT's first step into foreign direct investment policy. And finally, the Uruguay Round sought to accelerate the GATT process of trade dispute settlement, curb the abuse of antidumping laws, codify domestic subsidies and countervailing duties, and codify allowable safeguard measures.

These objectives, how they all fit into the grand bargain of the Uruguay Round and the trade successes achieved in the round, are more fully addressed in Chapter 7 on the new World Trade Organization. The primary achievement of the Uruguay Round was the creation on January 1, 1995, of the new World Trade Organization. With the authorization of the WTO, the world now had an institution for trade and investment comparable to the IMF and the World Bank for finance, and similar to the International Trade Organization which the U.S. Congress refused to ratify in 1946. Although many have often criticized various facets of GATT, the status of international trade that GATT handed off to the WTO in 1995 was vastly improved over what GATT had inherited when it was formed thirty-eight years earlier. It was an international trade system much more conducive to nurturing conspicuous customer solutions.

UNFAIR TRADE

In recent years, U.S. administrations have aggressively attacked "unfair trading practices" by U.S. trading partners, as covered more fully in Chapter 5. In essence, such actions have been a political response to the persistent U.S. multilateral trade deficits, and specifically to the stubborn U.S. bilateral trade deficit with Japan. However, a nation's overall trade balance is primarily a reflection of its domestic macroeconomic policies. When aggregate public and private expenditures exceed a nation's GDP, the difference must show as a current account deficit.

Correcting such imbalances primarily involves the domestic macroeconomic tools of fiscal, monetary, and currency exchange-rate policy more than specific trade measures. The idea of managing trade to achieve an improved trade balance not only runs counter to the market principles on which the U.S. economy has long been based, but is also counterproductive to the nurturing of conspicuous customer solutions.

If U.S. administrations officially bless the concept of managed trade, domestic industry after domestic industry will line up to garner its share of the benefits. We need only look at the last time the United States tried a similar approach. The Smoot-Hawley Act was an economic, social, and political disaster. Any international competitive advantage from managed trade will be temporary and will effectively undermine the rules-based system originally embodied in GATT and now resident in the WTO. A much better course would be for the United States to again take the lead in strengthening the multilateral rules and their enforcement, as the United States did in helping to realize the successful conclusion of the GATT Uruguay Round and the establishment of the WTO.

THE LEVEL PLAYING FIELD

A current lament of many U.S. politicians, trade unionists, and too many U.S. businessmen is for "fair" rather than "free" trade, and the right to play on a "level playing field." I can only assume that these people also believe in motherhood, apple pie, and the tooth fairy. In a business career spanning over fifty years, I have encountered few, if any, level playing fields, either domestic or foreign. And when I did, I worked just as hard as I could to convert the tilt in my direction.

The playing fields in foreign markets are always going to be tilted against invaders. Sometimes the barriers are blatant, sometimes they are more subtle. To penetrate a foreign market you must employ the economic discontinuities of innovation, variety, instant response, and quality, plus liberal doses of patience and persistence, often extending over a period of years. A one-week sales trip to a foreign market, during which you fail to land a single order, is not indicative of a closed market. It is the beginning of the marketing process.

Foreign markets will never be penetrated by new competitors who bring nothing more to the party than what is already present in that market. Some markets require bigger discontinuities, or more conspicuous customer solutions than others to penetrate.

The playing fields of Japan, France, and Germany are very un-even, but penetration is easier if you can speak the language.

An unfortunate belief of many in the U.S. business community is that the route to the realization of international trade opportunities runs through Washington, D.C. The successful penetration of specific foreign markets is not accomplished in your nation's capital, but through aggressive actions on the ground in the marketplaces of the world. Market to your potential customers, not to your elected officials and bureaucrats. And those who choose to await the creation of a level playing field before they enter the game will still be sitting on the sidelines when that game is long over.

In today's world of high expectations, there are few shareholders who will long tolerate a management who sits and waits for more favorable conditions to compete. So solve your own global market issues on the ground in those global markets. You can neither expect nor wait for your government to do your work for you. Government-induced solutions are normally short-term fixes that do not establish the type of long-term relationships needed to consistently convert global inputs into outputs for global markets just as efficiently and effectively as possible.

Chapter 5

Trade Distortions

As previously noted, the achievement of conspicuous customer solutions depends upon the complementary business tactics of international trade and foreign direct investment to provide broad access to world markets, and to provide minimal barriers to the ability to effectively convert global inputs into outputs for global markets. Trade distortions, the subject of this chapter, can best be described as actions taken by governments and/or enterprises that result in the altering of international trade flows from what they would otherwise be in an open, market-based, global economic environment. Such distortions reduce the ability to most effectively convert global inputs into outputs for global markets, and in the worst case, inefficiently utilize global resources.

Often called unfair trade practices, these trade distortions have deep roots in U.S. commercial policy, dating back at least from Alexander Hamilton's report on manufactures in 1791. However, the term *unfair trade* never appears in major international trade agreements, and although it does appear in the U.S. Trade and Competitiveness Act of 1988, it is not, therein, defined. In recent years, U.S. administrations have aggressively attacked unfair trade practices by U.S. trading partners. In essence, such action has been a political response to the persistent U.S. multilateral trade deficit, and specifically to the stubborn U.S. bilateral trade deficit with

Japan, even though it is broadly recognized that a nation's overall trade balance is primarily a reflection of its domestic macroeconomic policies and the global competitiveness of its private sector.

Although the list of potential unfair trade practices, some real and some imagined, is long, most such potential trade distortions can be included in a very few categories. First are actions by governments and/or enterprises that result in the export of unfairly priced (low-priced) products to other world markets. Included in this category are dumping, government subsidization of exports, and the granting of concessionary export credits.

Second are practices designed to create barriers that impede the access of exports to certain world markets, and include restraints such as import quotas, exclusionary technical, health, and safety standards, domestic purchasing preferences, and anticompetitive domestic practices such as *keirestu* in Japan.

The third category includes actions taken by national governments in support of their economic, social, political, or security agendas that have a negative effect on the ability of their domestic enterprises to export to certain world markets. These might well be considered self-inflicted trade distortions and include such actions as the imposition of strategic export controls, the imposition of economic sanctions, and the ex-territorial application of numerous domestic rules and regulations pertaining to taxes, competition policy, environmental protection, or labor policy.

Last, but certainly not least, is the troubling issue of trade distortions resulting from rampant public corruption and funded by bribes paid by private enterprises. Transparency International, an international good-government advocacy group based in Berlin, Germany, defines such corruption as the "misuse of public power for private benefit." Although the levels of such corruption appear highest in the developing world and corruption can be a strictly domestic activity, there is more than sufficient evidence that enterprises in the advanced economies are often the supply side of the bribery funds and that the payment of such bribes does distort normal trade patterns.

Unfair trade practices that cause significant distortions in international trade patterns can create massive pressures on the governments of individual nations to intervene in trade either to protect the home market of their domestic enterprises or assure access to foreign markets for their domestic exporters. However, when the only unfair practice is that the foreign competitor is more efficient or more innovative than his domestic counterparts, then the response of government intervention is counterproductive. Under this scenario of protectionist intervention, domestic resources are no longer optimally employed, and domestic consumers now pay higher prices, as do domestic producers, who use the product and are potential exporters. Such actions also create the potential for protectionist retaliation by trade partners and, at the same time, relieve the pressure on domestic enterprises to become more globally competitive and adjust to the new global realities.

DUMPING AND EXPORT SUBSIDIZATION

Long before the creation of GATT or the WTO, dumping and export subsidization were banned by many world nations. The United States authored countervailing duties to offset subsidies in 1897, and a U.S. antidumping law was enacted in 1916. The two practices are similar in that both involve selling into foreign markets at prices deemed to be at "less than fair market value." The practices are dissimilar in that in the case of dumping, the financial impact of the "less than fair market value" is absorbed by the exporter, whereas in export subsidization such activity enjoys some level of government financial support. Since domestic consumers could potentially enjoy the economic benefits of products priced at "less than fair market value," why should domestic governments be concerned by such imports? Fundamentally, dumping and subsidization are inconsistent with the GATT/WTO principle that international trade flows be based on market forces and not on government or other interventions. Such activity fails the perception of fairness, and unless viewed as fair, nations with market economies will not long support an open world-trade regime.

Initially GATT, and now the WTO, banned subsidies on manufactured products and dumping was condemned. In the GATT Uruguay Round, the subsidization of agricultural exports was also addressed, and this will be further discussed in Chapter 7, on the World Trade Organization. However, the right of an importing country to impose countervailing duties, as an import protection against dumping or subsidization, remains contingent on the additional finding of material injury to a domestic industry. Without such material injury, consumers may enjoy the resulting lower prices.

DUMPING

The GATT/WTO defines dumping as selling abroad at "less than fair market value." The determination of less than fair market is a price that is below that of the like product when the product is sold in the home market of the exporter. If the exporter does not sell the same product in his home country, in the absence of a comparable domestic price, the price at which the product is sold in other world markets may be used. If neither a home country nor other world market price is available, the standard used is the "constructed cost" of production in the country of origin, plus reasonable additions for selling cost and profit. Increasingly, the "constructed cost" has become the U.S. test of preference in prosecuting dumping cases.

Since 1980, approximately half of all U.S. antidumping cases have used this methodology. The U.S. Trade and Competitiveness Act of 1988 mandates the use of the "constructed cost" method for all imports from nonmarket economies. This recent proliferation of U.S. antidumping cases using the "constructed cost" standard suggests that this standard may have become a back door to achieving protectionism. Many U.S. trading partners believe that antidumping actions have become the tool of the new U.S. protectionists, and attempts were made during the Uruguay Round to curb the U.S. use of antidumping actions.

In many respects, the existing antidumping laws are blunt instruments designed for an earlier, more compartmentalized, global economic model. Predatory pricing to eliminate specific competitors is no longer a viable strategy in a world in which there are numerous capable global participants in almost every product segment. For a multitude of reasons, there are legitimate differences in product prices between different national markets and currency-exchange-rate movements can cause dumpinglike distortions in product prices between markets overnight.

Current antidumping rules should be revisited in light of global economic restructuring. Actual price comparisons should be virtually mandated when they are available, and when they are not and the constructed cost method must be used, average variable costs and not average total costs should become the measure of comparison. Similar to the U.S. Robinson-Patman Act, the meeting of competitive prices should be a justifiable defense in an antidumping case. All too often, in years past, highly protected markets have been the origin of low-priced products dumped in other world markets to achieve greater economies of scale. Thus, the ongoing process of market opening, through the removal of artificial barriers to trade, could also contribute to a reduction in the practices of dumping in a less cumbersome manner than lengthy antidumping dispute-settlement processes.

The good news is that in spite of all the attention dumping receives, it no longer significantly distorts trade flows. In the past two decades, antidumping laws affected no more than 0.5% of all U.S. merchandise imports, and less than 50% of the cases filed resulted in the imposition of duties.

GOVERNMENT SUBSIDIES

The United States is a party to the original GATT provisions on export subsidies and the 1979 Code on Subsidies and Countervailing Duties negotiated during the GATT Tokyo Round. In the GATT Uruguay Round, the United States sought to strengthen the code's disciplines on the use of subsidies, especially in global agricultural trade and by developing nations. Unfortunately, many other GATT members had a very different agenda for the Uruguay Round; that was to limit the excessive application by the United States of countervailing duty cases, with half of all the U.S. cases being brought against developing nations.

The GATT export subsidy rule, which the U.S. sought to strengthen, was a soft rule, and trade in agriculture was exempted from GATT overview. The soft rule did not specifically define banned export subsidies but did suggest some illustrative, but not all-inclusive, examples. In spite of the opposition to the U.S. objective, a significantly strengthened subsidy rule did evolve from the Uruguay Round, and trade in agricultural products, for the first time, was included in the overview of the GATT/WTO.

The GATT panel on export subsidies developed an innovative set of export subsidy classifications, identifying each with the colors normally associated with a traffic signal: red (prohibited), yellow (actionable), and green (permitted). Included in the prohibited category are domestic subsidies contingent on export performance or local value-added content. Remedies would permit other GATT members to impose unilateral countermeasures. In the actionable category are subsidies conferred on potential export industries and not made generally available throughout the economy. Tests here would include the percentage of exports in subsidized company revenues, the relationship of the subsidy to total product value, and whether the subsidy effectively distorts import or export patterns. The permitted category includes subsidies generally available to all enterprises, such as regional development assistance, the promotion of environmental protection, labor adjustment assistance, and the promotion of research and development.

As noted earlier in Chapter 3, the decision of the GATT panel on export subsidies to include the promotion of research and development as a permitted subsidy was quite interesting in view of the previous furor over the alleged R&D subsidy to U.S. enterprises from U.S. defense and space programs. It was, however, a pragmatic call and reflective of the important role that new ideas and innovation play in the process of economic development through their contribution to the creation of conspicuous customer solutions.

CONCESSIONARY EXPORT CREDITS

GATT's pre–Uruguay Round's "illustrative list" of banned export subsidies included concessionary export credits, which are credits offered usually on sales to the nations of the developing world at below-market interest rates and for longer than normal market repayment periods. Such credits frequently combine the public and private sector interest through what is called tied-aid and/or mixed credits, more fully described below.

However, the GATT Code effectively withdrew the GATT from implementing export credit policy through the transfer of this role to the OECD. Since the preponderance of all concessionary export credits are offered by the relatively few world nations that comprise the membership of the OECD, it was reasonable to assume that this role could be more effectively and efficiently performed by these fewer nations of more comparable economic circumstances.

Fundamentally, it is economically irrational to grant developing-world, higher-risk buyers more favorable credit terms than more credit-worthy domestic customers. The normal rational given is that such credits are a form of economic aid to the developing world. A closer review of the practice would suggest that perhaps export promotion, not economic aid, is the real driving force. This becomes more apparent in the case of tied-aid in which the economic aid is extended to a developing nation on the condition that the funds be spent buying product from enterprises in the aid-granting nation, or in the case of mixed credits, when part of the purchase price is funded by a government grant of economic aid and the balance is funded through a concessionary credit offered by the supplier.

The Development Assistance Committee of the OECD has long sought to dismantle bilateral tied-aid without much success. In 1996, the combined export credit exposure of all developing nations and countries in transition totaled just less than U.S.$500 billion, or almost one-fourth of the total indebtedness of these countries. Bilateral aid grants were U.S.$70 billion, almost half of which was tied-aid. Obviously, the interest of the aid-recipient nations could be better served and the cause of increased global competition aided if more of the economic aid to the developing world flowed through the international institutions established to provide this function, and higher levels of resources were committed to these institutions by the advanced economies, most of whom are OECD members. However, since, in the near term this does not seem to be a reasonable expectation, the next best solution would be an agreement on export credits between those nations that are the primary source of such credits. The OECD currently has such an agreement in place, and export credit practices that conform to the OECD Agreement on Export Credits are regarded as meeting the GATT/WTO rule.

The OECD Agreement on Export Credits establishes a schedule of minimum interest rates and maximum repayment periods for different categories of borrowers based on per capita national incomes, with the highest degrees of concessionality for the poorest countries. The agreement also provides guidelines for disciplining the use of mixed credits, with the maximum percentages of aid-grant applicable to the poorest countries.

RESTRICTIONS ON ACCESS TO FOREIGN MARKETS

Beginning in the 1980s, the emphasis of U.S. trade policy began to shift from combating unfair imports, such as dumped or subsidized products into the United States, to the pursuit of more open foreign markets for U.S. exports. This policy shift was very supportive of the emerging global paradigm of converting global inputs into outputs for global markets as efficiently as possible. The statutory basis for such action was Section 301 of the U.S. Trade Act of 1974, which authorized the president to attack such market restrictions through actual or threatened restrictions on access to the U.S. market. The scope of Section 301 authorizes the U.S. administration to unilaterally take action against unjustifiable, unreasonable, and discriminatory trade practices.

As defined in Section 301, unjustifiable foreign trade practices are those in violation of the international legal rights of the United States, and the United States has a strong legal position in this regard. Unreasonable practices are those not necessarily in violation of U.S. legal rights but that, nevertheless, are unilaterally deemed unfair and inequitable by the United States (i.e., export targeting, denial of worker or human rights, etc.). As regards unreasonable practices, the U.S. legal position is based on a U.S. unilateral finding and is significantly weaker than in the case of unjustifiable practices. Discriminatory foreign-trade practices are those such as the denial of national or most-favored-nation treatment for U.S. goods, services, or investment, normally available to all GATT members from all other GATT members.

The theory behind Section 301 is that the U.S. has the right to use access to its domestic market as leverage to open foreign markets and benefit all world exporters. Section 301 gives the U.S. administration broad powers to act against foreign-market practices the U.S. unilaterally judges to be unjustifiable, unreasonable, or discriminatory. If the U.S. threats under Section 301 result in successful bilateral negotiations to resolve the alleged unfair trade practice, then Section 301 has well served its intent.

But if these Section 301 threats fail and the United States chooses to unilaterally impose market sanctions in retaliation against practices not necessarily in violation of U.S. legal rights, the major deficiency and inherent danger in Section 301 becomes apparent. The United States is then operating in violation of the multilateral trading rules that it has worked so hard and so long to create, and is potentially liable for sanctions to which it has agreed for such forms of conduct.

In commenting on Section 301, a former vice minister of Japan's Ministry of International Trade and Investment (MITI) once said: "How can the United States, the champion of the GATT, of the WTO and of multilateral trade rules, invoke a system under which the United States unilaterally invokes its own criteria to determine unfairness, prosecutes its own case, and then hands downs its own sentence?" The United States needs to rethink the inherent dangers in a policy of self-determined retaliation as embraced by Section 301, which can potentially encourage a cycle of counter-retaliation and the erosion of a system of international trade rules, which the United States has spent fifty years fostering. The role for Section 301 has obviously been substantially reduced by the new WTO-accelerated and more enforceable trade-dispute-settlement process, and by the anticipated near-term inclusion of important U.S. trading partners, such as China and Russia, in the membership of the WTO. The determination of equity in a trade dispute made by a multilateral body, such as the WTO, is a much more credible solution than a U.S. unilateral trade decision.

The agenda before the dispute-settlement body of the WTO is heavy and growing rapidly. However, this function is one of the WTO's most important responsibilities, and properly managed, it has the potential to significantly reduce the number of trade distortions that have plagued the world trading system since the end of World War II.

STRATEGIC EXPORT CONTROLS

Turning from trade distortion caused by foreign governments and/or enterprises, there are also trade distortions created for exporters by actions taken by their own domestic governments in pursuit of that government's economic, political, social, or security agenda—often called self-inflicted trade distortions. The first of these are strategic export controls.

Ever since the early post–World War II era, the United States and its allies have tried to maintain rigid controls on the then communist regimes to limit those nation's access to critical military technology. The principal control mechanism was the Coordinating Committee of Multilateral Export Controls (COCOM), which was established in Paris in 1949. Despite numerous multilateral initiatives, the COCOM members were never able to unanimously agree on the list of restricted products, and the enforcement of the controls was very uneven between COCOM members. The United States, which without doubt was the largest potential supplier of such products, sustained the most vigorous enforcement of COCOM and had the longest list of restricted products.

In spite of its flaws, the COCOM system did succeed in denying or delaying the flow of many sensitive technologies to communist countries in years past. However, with the globalization of production and markets, plus the rapid dissemination of technology, including to non-COCOM member countries, the stemming of technology flow to the former communist countries became increasingly difficult. Even if there had been a perfect consensus within COCOM, ultimately there was sufficient expertise outside of COCOM to make strategic containment difficult.

The pace of technological change also made it difficult for U.S. and other global policy makers to keep up on new relevant developments and to maintain consistent controls under changing technological realities. U.S. strategic export controls were burdened with a scope that encompassed too many products and too many technologies to be administered effectively. At one time, the U.S. banned the export of numerous medical diagnostic products solely on the basis that the products contained an advanced integrated circuit.

U.S. controls also lacked a "sunset clause" to delist technologies that were no longer unique or products that were readily available from other foreign sources. U.S. export licensing procedures often discouraged potential foreign buyers through delays and uncertainties in the process that encouraged those buyers to seek other sources. At one time, in order to apply for an export license, the

U.S. exporter had to already have the order in his possession. You can imagine the reaction of the buyer when a U.S. exporter, after an extended delay, had to hand the order back because he was refused an export license. The complications of the export licensing process also discouraged many small U.S. firms, which lacked the necessary bureaucratic expertise, from seeking export opportunities.

Obviously, COCOM most heavily impacted high-tech industries that are burdened with high front-end research and development expenses. The more rigid U.S. enforcement of COCOM denied U.S. high-tech enterprises access to many markets willingly served by other COCOM members, thus making it necessary for these U.S. firms to amortize their R&D expenses on sales to fewer markets. The intent of COCOM was to deny certain countries access to strategically sensitive products and technologies. As noted earlier, COCOM was reasonably successful in either denying or delaying access to these countries. However, because of the manner in which COCOM was structured and the manner in which it was enforced by the United States, the trade distortion suffered by U.S. high-tech exporters was even greater and resulted in the relinquishing of certain world markets to foreign suppliers, whose government's enforcement of COCOM was less rigid.

As time progressed, it became apparent that there was an inherent conflict in strategic export controls as applied to dual-use technologies such as advanced communication and computer systems. Although these technologies had great military value, they were also important in aiding developing nations and nations in transition to modernize their economies and achieve a level of competitiveness consistent with economic self-sufficiency. Thus, the members of COCOM were forced to decide if the benefits of access to these technologies better served global interests by raising world living standards than the negatives resulting from the increased military capabilities created.

The members of COCOM wisely decided that the complicated COCOM regulations no longer effectively served their purpose, and that for world peace, higher national living standards and the process of inclusion made more sense than exclusion. A great deal of the former COCOM apparatus and the companion U.S. strategic export controls have now been dismantled. In their place is a new global arms-trade policy in which higher walls are being sought around a narrower set of technologies, particularly those applying to weapons of mass destruction and terrorism.

Under the terms of the new Wassenaar Agreement, thirty-three arms-exporting nations have agreed to a new set of controls relating to conventional weapons and to dual-use goods and technologies for export to the so-called pariah states of Iran, Iraq, Libya, and North Korea. Much of what has agitated the U.S. export community about trade distortions resulting from strategic export controls has now been dismantled. However, potential customers for U.S. exports, who were driven to foreign vendors by the former rigid U.S. strategic export controls, will not necessarily revert back to U.S. suppliers just because the U.S. controls have been relaxed. These potential customers now have current suppliers and must be won back to U.S. sources of supply by innovation, quality, price, and conspicuous customer solutions. Even though COCOM is gone, there are still global export and re-export controls on weapons and weapons-related technologies to certain world markets. These are strict regulations with harsh penalties.

ECONOMIC SANCTIONS

Throughout history, national governments have deliberately suspended normal trade and financial transactions with other nations to coerce actions they favor or actions they oppose. Recent examples are sanctions applied to Iraq, Iran, North Korea, the People's Republic of China, South Africa, Libya, and Haiti. Since the end of World War II, the United States has been the dominant user of economic sanctions. From 1945 through 1990, there were 107 sanctions imposed worldwide. The United States was involved in 69% of the 107, and U.S. unilateral sanctions constituted 52% of all sanctions imposed. Unfortunately, as the U.S. reliance on economic sanctions as a foreign policy tool has increased, their effectiveness has diminished. Many believe that such sanctions have little or no impact on the specific behavior of the countries subject to the punitive measure, in that in today's global economy a buyer that is refused service in one country can almost always find a willing seller in another. Thus, to achieve a major impact with unilateral sanctions, the imposing country must hold a near monopoly on trade with its target or be the exclusive supplier of a critical commodity.

Particularly harmful to U.S. enterprises have been unilateral economic sanctions that have exterritorial implications, such as recent initiatives to punish foreign entities engaged in certain commercial activities in Cuba and Iran. These measures antagonize U.S. trading partners, and the predictable results of such encroachments upon the sovereignty of other countries is that they adopt laws to block and counteract the sanction. Similarly, retaliatory countermeasures by these countries are not unusual.

Alternately, the sanction-imposer must achieve a high degree of cooperation among its allies, as has been the case in Iraq, South Africa, and North Korea. To achieve such cooperation, the interests of the allies must coincide. But the interests of the allies increasingly do not coincide to the extent required. In an interdependent global economy, poorly conceived sanctions not only fail their objectives but may cause more harm at home than abroad. In 1996, the President's Export Council (PEC) estimated that in 1995 the value of lost U.S. exports due to the imposition of economic sanctions to be U.S.$15 to $19 billion, affecting 200,000 to 250,000 export-related jobs. In the view of the PEC, the most important economic impact of unilateral economic sanctions is the cumulative weakening of U.S. competitiveness in friendly third-country markets, including those of our largest trading partners. Such indirect effects can include: special advantages created for foreign competitors in world markets; uncertainties about the availability of U.S.-originated goods, services, and technology; the unreliability of U.S. firms and their affiliates as suppliers and business partners, and retaliation by third-country governments and trading partners against interference in their international market decisions.

In view of the growing concerns over the negative economic impact of U.S. economic sanctions, particularly in the case of unilateral sanctions and their exterritorial application, in 1998 the U.S. Congress and the administration began to explore whether the U.S. use of sanctions is appropriate, coherent, and designed to gain international support. Among the changes under consideration are a cost-benefit analysis for proposed new sanctions, an annual "sunset review" to determine if existing sanctions should continue, and a narrowing of the sanction's targets to minimize hurting innocent victims, such as exempting humanitarian aid and programs of food and medicine. To this list should probably be added a review of the effectiveness of U.S. unilateral economic sanctions for which the United States is unable to solicit multinational support.

CORRUPTION AND BRIBERY

If you are seeking evidence that the world of international commerce is less than perfect, you need to look no further than the still-active role of corruption and bribery in many nations. Both Transparency International and the World Bank describe corruption as the "misuse of public power for private benefit."

There is no firm evidence that corruption and bribery are more rampant today than they have been in the past. However, numerous explanations can be advanced for the increased attention currently being paid to corrupt practices: the ending of the cold war, which destroyed the legitimacy of governments that were merely anticommunist; the transition from communism in Eastern Europe and the former Soviet Union, which revealed the extent of the corruption and provided opportunities for more; globalization, which has increased opportunities for corruption and placed private enterprises in contact with corrupt regimes; the rise of democracy; and the consensus on the superiority of the market, which has turned attention toward the role and functioning of the state.

Each year, Transparency International, a Berlin-based good-government advocacy group, publishes a Corruption Perception Index of nations. The group doesn't claim to be able to judge actual levels of corruption, but instead relies on the perception of corruption by the general public, business executives, and financial-risk analysts in establishing its ratings. The group's 1998 index is shown in Figure 5.1. A score of 10 is a corruption-free nation.

The obvious conclusion from a review of the Transparency International Index is that corruption and bribery are perceived to be more pervasive in the developing world and in the countries in transition than in the advanced economies. In these less-than-full market economies, the centrally administered governments rely on a myriad of detailed laws and regulations, which form the

Figure 5.1
Transparency International Corruption Perception Index, 1998

Risk	Country	Score	Risk	Country	Score	Risk	Country	Score
1	Denmark	10.0	29	Malaysia	5.3	55	Senegal	3.3
2	Finland	9.6		Namibia	5.3	59	Ivory Coast	3.1
3	Sweden	9.5		Taiwan	5.3		Guatemala	3.1
4	New Zealand	9.4	32	South Africa	5.2	61	Argentina	3.0
5	Iceland	9.3	33	Hungary	5.0		Nicaragua	3.0
6	Canada	9.2		Mauritius	5.0		Romania	3.0
7	Singapore	9.1		Tunisia	5.0		Thailand	3.0
8	Netherlands	9.0	36	Greece	4.9		Yugoslavia	3.0
	Norway	9.0	37	Czech Rep.	4.8	66	Bulgaria	2.9
10	Switzerland	8.9	38	Jordan	4.7		Egypt	2.9
11	Australia	8.7	39	Italy	4.6		India	2.9
	Luxembourg	8.7		Poland	4.6	69	Bolivia	2.8
	United Kingdom	8.7	41	Peru	4.5		Ukraine	2.8
14	Ireland	8.2	42	Uruguay	4.3	71	Latvia	2.7
15	Germany	7.9	43	South Korea	4.2		Pakistan	2.7
16	Hong Kong	7.8		Zimbabwe	4.2	73	Uganda	2.6
17	Austria	7.5	45	Malawi	4.1	74	Kenya	2.5
	United States	7.5	46	Brazil	4.0		Vietnam	2.5
19	Israel	7.1	47	Belarus	3.9	76	Russia	2.4
20	Chile	6.8		Slovak Rep.	3.9	77	Ecuador	2.3
21	France	6.7	49	Jamaica	3.8		Venezuela	2.3
22	Portugal	6.5	50	Morocco	3.7	79	Columbia	2.2
23	Botswana	6.1	51	El Salvador	3.6	80	Indonesia	2.0
	Spain	6.1	52	China	3.5	81	Nigeria	1.9
25	Japan	5.8		Zambia	3.5		Tanzania	1.9
26	Estonia	5.7	54	Turkey	3.4	83	Honduras	1.7
27	Costa Rica	5.6	55	Ghana	3.3	84	Paraguay	1.5
28	Belgium	5.4		Mexico	3.3	85	Cameroon	1.4
				Philippines	3.3			

multiple opportunities for bribes to circumvent. Additionally, the pay of low-level government employees in the developing world is often low, and these employees rely on what is known as "grease payments" for small favors to supplement their incomes. Sales to government agencies, state enterprises, or sales requiring government approval are particularly prone to bribery.

However, this is not to say that the advanced economies are immune to bribes, particularly involving sales to government agencies or government-owned enterprises. In recent decades, the revelation of bribes paid on the sale of American-manufactured aircraft to government-owned airlines in the Netherlands and Japan soiled the reputation of Prince Bernhart of the Netherlands and toppled the government of Prime Minister Tanaka of Japan.

Furthermore, the Transparency International Index also raises the disturbing question as to the source of the bribes paid in the developing world. In too many instances, the supply side of the bribe transaction is provided by enterprises based in one of the advanced economies. Thus, corruption is not just a developing-world domestic practice, but an international virus that has the ability to distort normal trade flows.

It was, in fact, the Tanaka incident in Japan that caused the U.S. Congress in 1977 to enact the now infamous Foreign Corrupt Practices Act (FCPA), which has struck paralyzing fear into the hearts of most U.S. enterprises and businessmen engaged in international commerce. When enacted, the FCPA made the United States the only country in the world that banned its enterprises from trying to subvert the political processes of a foreign country and buy the services of a politician or administrator to ease its way into a market. Only the United States had legislation that made bribery of foreign officials illegal; in France and Germany, foreign bribes were tax deductible as a normal cost of doing business.

There were many causes for this reluctance to keep companies from doing business in foreign capitals that they were prohibited from doing at home. High-minded types spoke of an unwillingness to legislate their morality beyond national borders. Recipients of the corporate largesse cloaked their gains in the language of local cultural particularities: the necessity of gift giving has long been a favored justification and explanation. Business officials from the advanced nations were happy to exploit the competitive advantage they had over U.S. competitors, which resulted in the distortion of trade flows and the loss to U.S. companies of over one hundred business deals, estimated at over U.S.$50 billion. As we shall see, in spite of heavy U.S. pressure, it was not until 1998 that the other members of the OECD agreed to follow the lead of the United States, as expressed in the FCPA.

Although the goal of the FCPA was laudable, at the time of its enactment it inhibited the actions of only U.S. enterprises, and the ambiguity of the act's language, its application and intent, scared off or hampered many U.S. businesses from competing for legitimate international business opportunities due to the act's harsh civil and criminal penalties. Following passage of the act, Congress began to realize some of the harm that had been done, and in 1988 the Trade and Competitiveness Act attempted to rectify some of the earlier damage by tightening up definitions, clarifying accounting issues, and expediting government responses to information requests.

A major improvement to the legislation was the establishment of the difference between a bribe, an illegal act, and a facilitating payment, a legal act. A facilitating payment, also known as grease, was defined as a payment to expedite an action to which you were legally entitled, such as the issuance of a driver's license, an inspection of equipment, or to accelerate the payment of an invoice. A bribe was defined as a payment made to influence a decision in a matter to which you are not entitled by right, such as the award of a competitive contract.

The reimbursement of travel expenses for foreign government officials was deemed legal as long as they were directly related to the promotion, demonstration, or explanation of products or services, or in conjunction with the performance of a specific contract, and were generally in conformity with normal commercial practice. Because the FCPA bans both direct and third-party bribes, the 1988 revisions further defined the act's reason-to-know standard as regards the conduct of sales agents, often involved in these types of transactions.

Finally, the act recognized the good-faith efforts by a minority shareholder in a joint enterprise to cause the adoption of accounting controls consistent with the FCPA. The use of the local branch of a U.S. audit firm was considered good-faith compliance. Further obligations of a U.S. shareholder increased as his percentage of ownership increased or his active participation in management or board activities increased.

In May 1996, after years of intense pressure from the United States, a ministerial meeting of the OECD agreed to "criminalize the bribery of foreign officials in an effective and coordinated manner" and reexamine the tax deductibility of bribes to foreign officials, where this was still permitted. In May 1997, the OECD published a "good-behavior undertaking," which encompassed most of its 1996 objectives. The role of the OECD in this situation is extraordinarily important, in that the supply side of the foreign bribe is frequently a private enterprise whose home government is an OECD member.

In Paris on November 22, 1997, the twenty-nine countries that belong to the OECD formally agreed to a treaty that would outlaw the bribery of foreign officials. The treaty marks an important victory for the United States, which took such action in 1977. Although the OECD has no power to insure compliance with the treaty, the signatories are required to monitor themselves, as well as each other, and to ensure that the sanctions against bribery imposed by each country are sufficient. The nations agreed to present the agreement to their parliaments by April 1998, and hoped to change their individual laws to make the agreement effective by the end of 1998. Once ratified by the individual governments, the treaty will become a legally binding agreement. By mid-1999, fifteen member nations had advised the OECD of such ratification, with the balance expected by year end.

As noted at the beginning of this chapter, the list of potential trade distortions is long, and such transactions have significantly impacted global trade flows in the past. Although the world will never be without trade distortions, actions by GATT and the WTO have effectively addressed many of these issues, domestic governments are effectively addressing many others, and the OECD has effectively addressed the virus of bribery. The net result is a more open, market-based global economy with an enhanced ability to efficiently convert global inputs into outputs for global markets and to create conspicuous customer solutions.

Chapter 6

Multilateral versus Alternative Trade Practices

For the first thirty-five years following the end of World War II, the United States pursued an international trade policy firmly based on the GATT principles of multilateralism and nondiscrimination. As used in this context, multilateralism describes the practice of taking a collective or universal approach to negotiating trade arrangements, rather than negotiating them on a country-by-country or regional basis. Nondiscrimination, in turn, describes the policy of treating all trading partners equally, as mandated in the most favored nation (MFN) policy contained in the GATT treaty.

During this period, the United States was the main advocate of trade arrangements based on these principles and the main opponent of discriminatory trade schemes. Both U.S. policy and the GATT trade principles were based on the widely held presumption that such conduct was most likely to provide the freedom of market access necessary to encourage the creation of conspicuous customer solutions and the ability to effectively convert global inputs into outputs for global markets.

However, in more recent years, U.S. exceptions to multilateralism began to grow in both number and importance. Primarily, these U.S. exceptions were bilateral, sectorial agreements designed to restrict foreign access to U.S. markets for products such as steel (voluntary export restraints), automobiles (a quota on the importation of cars from Japan), and clothing and textiles (the Multifibre Agreement, or MFA). Other U.S. exceptions had trade-liberalizing objectives, such as the Free Trade Agreement (FTA) with Canada and its subsequent extension to NAFTA, which included Mexico and perhaps other Latin American nations. Even though these free-trade agreements were trade liberalizing in intent, they were essentially bilateral and regional, but not multilateral, and discriminated against those not included.

This growing U.S. inclination to pursue trade liberalization outside of the traditional pattern of multilateralism was primarily a domestic political reaction to the growing U.S. frustrations regarding its endemic merchandise trade deficits. These frustrations were fueled by the perception that other trade partners did not play by the same rules as the United States, the perception that the domestic economies of some major trading partners did not act in conformity with the implicit market economy assumptions on which the multilateral rules were based (*keiretsu* in Japan), and the perception that the GATT dispute-settlement process was slow and at times unenforceable. There were also frustrations that GATT rules had not evolved to include the new areas of trade of keen interest to the United States, such as trade in services, intellectual property rights, and trade-related investment measures. Although coverage of these issues was ultimately achieved in the GATT Uruguay Round, as was significant improvement in the trade-dispute-settlement apparatus, these solutions were late in coming.

At this same time, some Americans began to advocate the bilateral negotiation of specific trade outcomes in terms of volumes or market shares, even though such a policy of managed trade would fail the three most important principles of an effective trade policy (i.e., fostering specialization, encouraging competition, and aiding economic adjustment). In many ways, this heavy-handed approach to trade policy was similar to the thought process that produced the disastrous Smoot-Hawley Act of 1930. Nevertheless, the persistence of these U.S. frustrations, the successful negotiation of NAFTA, and the obvious successes of the European Union prompted some compelling questions.

Should the United States pursue additional discriminatory trade agreements? Should the pursuit of such arrangements remain the exception to U.S. trade policy, or should they become the rule?

What are the implications both for the United States and the world trading system of such a U.S. policy change, and where did U.S. interests lie? The conventional wisdom had always been that U.S. interests were best served by preserving multilateralism although special circumstances might arise, such as in the NAFTA case, in which exceptions were warranted. One of the most robust conclusions to come out of the study of economics in the past two hundred years is the general presumption in favor of freer trade on a MFN basis.

Following the end of World War II, GATT established world trade rules, resolved trade disputes, and was the instrument through which the liberalization of world trade had been negotiated. Historically, GATT had experienced serious shortcomings both as a rule book and as a referee, but the GATT Uruguay Round and the new WTO specifically addressed these issues and provided more powerful solutions. However, these earlier GATT shortcomings appeared to some as the reason that multilateralism might no longer be appropriate.

It should be remembered that GATT was a creature of its own members, no more and no less. Its most serious problem had been a lack of will on the part of its members to make it work. The Uruguay Round strengthened and accelerated many GATT procedures and deeded to the new WTO greater enforcement powers. Therefore, it was most logical to assume that the interests of the United States would be best served by supporting the WTO framework to make the world trading system work. In particular, it was necessary to consider the consequences of adoption by the United States of trade practices outside of those for which it had worked so hard to have included in the WTO structure.

THE TRADITION OF MULTILATERALISM AND NONDISCRIMINATION

George Santayana once warned that those who cannot remember the past are condemned to relive it—and who can forget Smoot-Hawley? When the authors of GATT adopted the unconditional most favored nation provision, they were applying lessons learned from harsh experience. Discriminatory trade agreements had been a contributor to the tensions of the Great Depression that helped to trigger World War II. The most favored nation principle was imbedded in paragraph 1 of Article I of the GATT agreement, and clearly stated that "any advantage, favor, privilege, or immunity granted by any contracting partner . . . shall be accorded immediately and unconditionally to . . . all other contracting parties." Multilateralism and nondiscrimination also became the foundations for U.S. trade policy.

The GATT experience strongly supported the proposition that freer trade on a nondiscriminatory basis promotes economic growth and improves global welfare. The most obvious direct result of GATT's multilateral trade negotiations had been the reduction of tariffs to the point that, in general, tariffs no longer constituted a significant trade barrier between the advanced economies. By generalizing the results of trade liberalization to all GATT members, greater liberalization had occurred and trade expansion had been accelerated. It had also made possible full participation in the world economy by a growing number of nations.

In many ways, GATT was a victim of its own success. Growing numbers of participants and progressive tariff reductions created new problems with which its member nations enjoyed a diminishing ability to successfully solve. With more participants, the process increasingly encountered the "free rider" and "convoy laggard" problems. During a multilateral negotiation, when concessions are to be extended to all parties, there is a political incentive to hold back, keeping one's own tariffs up while hoping to get the advantage of other countries' tariff reductions. Therefore, the only way to avoid free riding is to make cuts only when each participant is willing to so do. Thus, the least-willing participant determines the pace of the negotiations, and the speed of the convoy moving toward freer trade is limited by the speed of the slowest ship. For a large country such as the United States this can be particularly frustrating, because it has little opportunity to hitch a free ride. Unless the United States goes along, a general cut in trade barriers is unlikely.

The growth in GATT membership, from twenty-three nations at the time of its creation in 1948 to over 120 nations by the time of the Uruguay Round, made consensus more difficult to achieve. Each of the eight GATT negotiating rounds took longer, with the last round, the Uruguay Round, taking seven years to accomplish. Participation by countries at dramatically different stages of economic development also made the negotiating problem more difficult. Many nations of the developing world opposed expansion of the scope of the multilateral rules and became convoy laggards. And finally, as tariffs declined and trade expanded, economic adjustment costs had grown higher, with the result that many nations sought to avoid actions that further increased the cost of economic dislocation. In certain cases, nontariff barriers rose to replace former tariff barriers.

By 1982, the convoy laggard problem had worsened to the point that it paralyzed the ability to further pursue multilateral negotiations. At that time, a special GATT ministerial meeting failed to agree on the agenda for a new GATT negotiating round. The last previous round had been the Tokyo Round, which concluded in 1979. This failure, plus growing political pressure for protectionism in the United States, prompted the United States to initiate a search for alternatives to multilateral GATT mechanisms. Indications that the United States was prepared to explore GATT alternatives fell on fertile ground. Two U.S. trading partners with major stakes in access to the U.S. market, Canada and Israel, initiated free-trade agreement negotiations with the United States.

Although multilateralism and nondiscrimination had guided international trade for over forty years, the architects of GATT had originally recognized that potential benefits lay outside of the MFN mainstream if certain conditions were met, and had made provision for them in the GATT agreement. Some of these exceptions are trade liberalizing in intent, such as economic cooperations (the Asia-Pacific Economic Cooperation), free-trade areas (NAFTA), customs unions (the European Union), and plurilateral agreements (the Tokyo Round Code on Government Procurement). Others are trade restricting in nature and include bilateral, industry-specific agreements, such as the MFA for trade in clothing and textiles, voluntary export restraints (VERs) for steel, and import quotas such as the quota on the importation of Japanese cars into the United States.

At the present time there are over sixty preferential trade agreements in existence, ranging all the way from economic cooperations, to trade preference associations, to free-trade areas, to customs unions, to common markets and economic unions. The two most common forms are free-trade areas and customs unions, with the European Union being the only association to have advanced to the ultimate classification of an economic union, with its adoption of the Euro as a common currency in 1999. A major difference between free-trade areas and customs unions is that although both completely eliminate barriers against imports from each other, the members of an FTA each maintain their own barriers against imports from others, whereas the members of a customs union maintain an identical common barrier against imports from others. A problem with the FTA rules is that these disparate barriers to the outside world, maintained by its separate members, require the adoption of complex rules of origin, by which the members must agree on minimum levels of internal value-added required for the product to move freely within the FTA, in order to prevent all importation into the FTA through the member country that offers the lowest level of external tariffs.

The acceleration toward the creation of regionalized trading schemes was fostered by the growing frustration with the GATT system, the perception that regionalism is here to stay (even the United States has embraced the concept), and that regional agreements were thought to be easier to negotiate.

THE CASE FOR REGIONAL TRADE ARRANGEMENTS

In their 1989 paper, "Is There a Case for Free Trade Areas," economists Paul Wonnacott and Mark Lutz make the case that free-trade agreements, which depart from multilateral, nondiscriminatory treatment, do not necessarily improve economic efficiency, raise real incomes, nor represent a move toward free trade. They note an FTA may be trade-creating, generating additional international trade and improving real incomes. But it is also a discriminatory arrangement. Some nations are in; some are out. Thus, trade may be rearranged and redirected when an FTA is formed. Such trade diversion generally represents a step away from economic efficiency.

The authors suggest the most appropriate manner in which to judge the overall economic effect of an FTA would be to compare real income under an FTA with the real income that would occur with the status quo. Since such a comprehensive approach is difficult, they suggest the distinction first proposed by the economist Jacob Viner in his 1950 classic, *The Customs Union Issue*, which regards the balance between trade creation and trade diversion as the core FTA efficiency test. Where new international trade is created by FTA members importing from other members what they used to produce domestically, efficiency is improved in that the imported products are less expensive. However, if FTA members now buy from other FTA members products they previously imported from outside countries, then efficiency generally falls in that the FTA member is generally a higher cost producer than the outside country, or the product would have been imported from the FTA member even before the FTA was formed. In practice, an FTA will have both trade-creating and trade-diverting effects. If trade creation dominates, there is the presumption that economic efficiency has been improved. If trade diversion dominates, Viner concluded that the effect would be a decline in efficiency.

The distinction between the trade-creating effects and the trade-diverting effects of an FTA provides powerful support for GATT Artical XXIV, which establishes the ground rules for participation in FTAs by GATT members. Article XXIV requires that restrictions be removed on substantially all of the trade between the parties to the agreement, as partial agreements are much more likely to produce trade diversion. It also requires this scope of coverage to be achieved within a reasonable amount of time pursuant to a specific plan, in that given the freedom to pick and choose, negotiators are more likely to choose products that result in trade diversion than trade creation. And finally, Article XXIV forbids barriers to nations outside of the agreement being raised as a result of any FTA arrangement. The raising of third-country barriers can only be motivated by the intent to divert trade to the FTA partners.

It is possible to postulate additions to Article XXIV that would further promote trade efficiency. Instead of mandating that barriers to trade with nations outside of the FTA not be raised, one could mandate that these barriers be significantly reduced. Another possibility would be to eliminate the FTA option and require all preferential arrangements to take the form of customs unions. This would eliminate the need for complex rules of origin, which are highly restrictive instruments of protection. And finally, Article XXIV could require that preferential arrangement members adopt liberal terms of accession permitting the inclusion of any nation willing to abide by its rules.

THE U.S. EXPERIENCE WITH EXCEPTIONS
TO MULTILATERALISM

The United States is currently a participant in a wide range of exceptions to the policy of multilateralism and nondiscrimination in international trade. It is a participant in two free-trade areas, that is, with NAFTA and Israel. It is a participant in the Asia/ Pacific and Enterprise for the Americas economic cooperations. It has numerous bilateral, sectorial agreements with over forty countries, covering an even larger number of industry segments. And, it is a participant in the Tokyo Round Plurilateral Conditional MFN agreement on government procurement. Unfortunately, this patchwork is the result of years of ad hoc political responses to special problems and opportunities, and not a considered product of a long-term trade strategy. The original FTA with Canada was the most ambitious and significant of the special arrangements into which the United States has entered.

The U.S.-Canada free-trade agreement was a case of two advanced economies, living in close proximity, with a long history of peaceful relations, including the longest undefended border in the world and a high degree of economic integration prior to the free-trade agreement. Each was and still is the other's largest trading partner, with more than 70% of Canada's exports coming to the United States.

The free-trade agreement was primarily Canadian driven. As a result of previous GATT multilateral tariff reductions, Canadian manufacturers no longer enjoyed a comfortably protected domestic market and needed to achieve the required economies of scale to participate in greater export markets, predominantly in the United States. In recent years, Canada has become worried about sudden changes in access to the U.S. market, and particularly about the frequent U.S. application of countervailing duties. Thus, fear rather than hope drove the Canadian initiative.

The U.S./Canada Free Trade Agreement did establish improved procedures for consultation and trade-dispute settlement, and did provide Canada some protection against the arbitrary extensions of U.S. laws. Thus far, it seems to be benefiting both nations, and both economies have become more efficient. Despite the former close relations between the United States and Canada, the negotiations were not easy, and nearly collapsed on several occasions.

The belief that regional agreements might be easier to negotiate did not prove to be the case in the U.S./Canada Free Trade Agreement.

In the end, Canada was forced to call a national election as a referendum on the free-trade agreement with the United States. The Progressive Conservative party of Prime Minister Brian Mulroney ran on the platform of favoring the agreement. The Liberal party and the New Democratic party ran on platforms opposing it. In the election, no party achieved a majority of the votes, but Prime Minister Mulroney's party did receive a plurality and the referendum was approved. However, it succeeded only because of the three-way split of the vote, and more Canadians actually voted against the FTA than voted for it. If held again today, the Canadian response would probably be much more favorable.

The U.S./Israel FTA is between countries of vastly different size, very different stages of economic development, and separated by great geographic distance. It is grounded in special historical, political, and strategic, but not trade, circumstances. To date, any commercial benefits have yet to be discernible. This FTA again raises the issue as to whether FTAs can be productive between countries of significantly different size, significantly different stages of economic development, and separated by wide geographical distance.

The United States is also a participant in a growing number of sectorial agreements designed to restrict trade or manage trade in such products as textiles, steel, machine tools, autos, and semiconductors. Authored in the name of correcting marketplace distortions, there is a basic similarity between these sectorial agreements. They are addictive, they retard inevitable adjustment to international competition, they impose large costs on the U.S. economy and the U.S. consumer, and, last but certainly not least, they offer no useful model for advancing U.S. interests in international trade. As noted earlier, they also fail the three basic dynamics in a successful trade policy: fostering specialization, encouraging competition, and aiding economic adjustment.

The Plurilateral, Conditional MFN agreement on government procurement entered into by the United States and certain other nations in the GATT Tokyo Round was an interesting attempt to open the government procurement market to international tender from countries that also opened their government procurement markets to international tender. The hope was that this conditional group would be expanded over time to be GATT-wide and to remove barriers in a market long restricted to domestic competition. Although this conditional MFN group is still less than GATT-wide, government procurement is much more open now than it was in 1979, and this plurilateral agreement has been of significant assistance to the new WTO in further prying open the government procurement markets.

The United States, although the main opponent of discriminatory trade agreements in the post–World War II period, vigorously supported the concept of the economic integration of Europe for political and strategic reasons and as a complement to the Marshall Plan. In the early years of the European Union's existence, a substantial volume of trade was diverted from outside countries as the consequence of the discrimination inherent in a customs union. Almost 40% of the increase in internal EU trade resulted from trade diversion rather than trade creation. However, as time passed and the common external tariffs were lowered through the actions of GATT, the benefits spilled over to trade with non-EU nations. Beginning in 1973, the EU's external trade, as a percentage of its gross domestic product (GDP), was greater than prior to the formation of the EU. Since trade diversion was then and continues to be overshadowed by trade creation, it must be assumed that the EU's economic efficiency, and probably that of the whole world, has increased.

Experience to date would indicate that multilateralism and nondiscrimination will not be the exclusive principles in the practice of international trade. There is, today, a parallel path of regional arrangements that can be either supportive or in opposition. The principles of multilateralism and nondiscrimination in international trade will be best served if regional arrangements become building blocks rather than stumbling blocks to an open, global system of trade and investment.

To the extent that multilateral trade negotiations accomplish a reduction in the tariff margins enjoyed by the regional blocs, the two paths of liberalization, global and regional, will tend to converge. Thus, the ultimate regional bloc could well become one that includes all trading nations, and with it the ability to efficiently convert global inputs into outputs for global markets.

Chapter 7

The World Trade Organization

As stated early on, the primary objective and principal benefit of international trade is the improvement of global economic welfare by making it easier to efficiently convert global inputs into outputs for global markets and to create conspicuous customer solutions. International trade contributes to this objective through the dynamics inherent in the process of international transfer: fostering specialization, encouraging competition, and aiding economic adjustment.

On a global basis, trade must be symmetrical, with total exports matching total imports. But, on a national basis, trade is seldom symmetrical and is reflected in surpluses or deficits in the current-account balances of individual nations. Nations with current account surpluses are net capital exporters, and nations with current-account deficits are net capital importers.

Earlier chapters have reviewed the history of international trade primarily from the end of World War II until the present. It was noted how in 1945, international trade and global economic integration had fallen to historically low levels, following a thirty-year period that included two world wars, the Great Depression and the aftershocks of the U.S. Smoot-Hawley Trade Act of 1930.

In this economic environment, the American and British architects of the post–World War II economy deliberately set out to reintegrate the global economy through the establishment of new international institutions. Included in these were the International Monetary Fund (IMF) and the International Bank for Reconstruction and Development (IBR&D), for finance. The IMF still exists today, as does the IBR&D, now called the World Bank. Also proposed was an International Trade Organization (ITO) to oversee global trade. When the U.S. Congress refused to ratify the concept of the ITO, it was replaced in 1947 by the General Agreement on Tariffs and Trade, which essentially consisted of the commercial policy chapter taken from the failed ITO. GATT remained in existence until 1995, when it was replaced by the World Trade Organization, which was conceived and authorized in the GATT Uruguay Round. During its lifetime, GATT was the vehicle by which multilateralism and nondiscrimination in trade were integrated into global trade policy.

Through GATT's first seven rounds of trade negotiations, including the Tokyo Round, which was concluded in 1979, the principal accomplishment of GATT was the reduction of tariffs on merchandise trade between the advanced economies of the world. GATT also promulgated trade rules regarding trade distortions, such as dumping, export subsidies, and unreasonable restrictions on access to foreign markets. Through the years of GATT's existence, it was the international-trade rule book and the adjudicator of trade disputes between nations. During this period, international trade grew twice as fast as world economic output, and trade was generally viewed as the engine of global economic growth.

As time progressed, certain nations also sought trade liberalization through measures that were less multilateral and more discriminatory in nature, such as regional trade arrangements, including free-trade areas and customs unions. Today, there are more than sixty such arrangements, with the European Union being the largest and most successful. There was also growth in bilateral, sectorial agreements designed to restrict trade.

The GATT rules, from the very beginning, provided for exceptions to nondiscrimination in trade to accommodate regional arrangements, if certain conditions were met. The intent of these exceptions and conditions, contained in GATT Article XXIV, was to encourage trade creation rather than trade diversion.

THE GATT URUGUAY ROUND

By 1986, GATT, conceived as an interim organization, was in its thirty-eighth year of existence and was beginning to encounter some severe problems. The "free rider" and "convoy laggard" syndromes were slowing the rate of trade liberalization. GATT rules had not been extended to new areas of trade of keen interest to many GATT members, such as trade in services, intellectual property right protection, trade in agriculture, and trade-related investment measures. The GATT dispute-settlement system was viewed as slow and weak. And there was growing divergence in views on trade policy between the advanced economies seeking greater and broader trade liberalization, and developing nations, experiencing greater difficulties in dealing with the economic adjustments resulting from earlier GATT-sponsored trade liberalization.

Although in 1982 a GATT ministerial meeting had failed to agree on the agenda for a new GATT negotiating round, under continuing pressure from the United States, the GATT members finally assembled in September 1986 in Punta del Este, Uruguay, to launch the Uruguay Round of Multilateral Trade Negotiations, which would turn out to be the eighth and last such round conducted under the auspices of GATT. The round, which took seven years to negotiate, was concluded in Marrakech, Morocco, in April 1994. It was, by any measure, the most ambitious negotiating round, and is generally regarded as a landmark achievement in providing great benefits to all trading nations.

The rationale behind the naming of this GATT round, and the fact that it was both initiated and concluded in nations of the developing world, was to emphasize the important developing-nation issues in the negotiation and to emphasize the grand bargain between the objectives of the developing nations and the objectives of the advanced economies on which the negotiating strategy was based.

The developing nations sought greater and more secure access for their agricultural and simple-manufactures exports to the markets of the advanced economies. Specifically, they sought lower tariff bindings, new rules for agricultural trade, new rules pertaining to trade in textiles and apparel, and new rules pertaining to the issuance of industry safeguard measures in the advanced economies. In return, the advanced economies sought developing-nation acceptance of new rules pertaining to trade in services, acceptance of new rules pertaining to intellectual-property-right protection, and the acceptance of new rules pertaining to trade-related investment measures. To best gauge the success of the Uruguay Round, it is instructive to review its specific accomplishments.

As noted earlier, IMF staff estimates that import-weighted, average, bound-tariff rates for the advanced economies, pre–Uruguay Round, were 6.0%, and average applied rates were 5.0%. Import-weighted average bound rates post–Uruguay Round for these same advanced economies were 3.6%, a 40% reduction in bound rates, compared to an objective of 33% for the round. In addition, advanced-economy bound rates on certain developing world exports of great importance, such as wood, pulp, furniture, paper, metals, and mineral products, were reduced by as much as 50% to 60%.

Trade in agriculture had never previously been included in the scope of GATT, and was a subject of great political and social stress in both the advanced economies and in the nations of the developing world. Many of the advanced economies, under political pressure from their farmers, subsidized domestic agriculture with high levels of support payments, which produced large domestic surpluses of high-priced product. Similarly, many of the advanced economies established quotas on categories of agricultural imports (such as Japan and Korea on rice) or charged variable levies (tariffs) to raise the prices of agricultural imports to domestic market levels (the Common Agricultural Policy of the European Union). To dispose of the agricultural surpluses generated by such support payments, many of the advanced economies subsidized the sale of these surpluses and dumped the products on the world market at distress prices. The agricultural exports from the nations of the developing world thus faced access restrictions to many of the markets of the advanced economies and subsidized competition from the advanced economies in other global agricultural markets.

Under the new rules for trade in agriculture agreed to in the Uruguay Round, all domestic agricultural support payments must be reduced by 20% by the advanced economies, and 13.3% by developing nations, during an implementation period of six years for the advanced economies and ten years for the developing countries. All agricultural export subsidies must be reduced by 36%, and the quantity of subsidized exports reduced by 21%, also during a six-year period. Since developing nations did not generally subsidize agricultural exports, this primarily impacted the advanced economies.

All existing import quotas on agricultural products were to be eliminated and replaced by tariffs. All tariffs on the import of agricultural products were to be reduced by 36% for the advanced economies and 24% for the developing nations over a six-year period for the advanced economies, and a ten-year period for the developing nations. And finally, rules pertaining to food and animal safety standards were to be harmonized to discourage their use as barriers to trade.

Prior to the Uruguay Round, trade in textiles and apparel had been the most highly protected area of trade in manufactured goods. Under the former Multifibre Agreement, all trade in textiles and apparel were subjected to a complex system of bilateral product-by-product quotas for each world nation. Under the new Uruguay Round rules, these quotas were to be eliminated and replaced by tariffs according to the following schedule: by January 1995, 17% of all imports were required to be converted to tariffs; by January 1998 another 17%; by January 2002 another 18%; and all remaining imports were to be converted to tariffs by 2005. When all quotas are converted to tariffs, trade in textiles and apparel will be subject to the same disciplines as all other sectors of merchandise trade, including negotiated tariff reductions.

GATT Article XIX allowed GATT members to take temporary "safeguard" actions to protect specific domestic industries from an unforeseen increase in imports that could cause serious injury to those industries and their workers. The theory was that this "safeguard" grace period would permit the impacted industries to either adapt to the new global competitive realities or shift assets into different activities in which they might better enjoy comparative advantage. The problem was that "temporary" was seldom "temporary," and once protected these industries had little incentive to take actions necessary to cope with the global realities and return to the status of global competitiveness. The new GATT agreement specifically prohibited so-called grey areas, or non-transparent actions designed to achieve protectionism.

The GATT agreement also established a "sunset clause" on all safeguard actions, under which all GATT Article XXIX safeguards must be terminated no later than eight years after first applied or five years from the date of establishment of the WTO. The maximum term for new safeguards was set at four years, and new safeguards had to be liberalized (their level of protection systematically reduced) each year after imposition. A GATT Safeguards Committee was established to oversee this area of trade and to provide surveillance of the safeguard commitments.

Based on the actions taken regarding bound-tariff rates on exports to the advanced economies, trade in agriculture, textiles, and apparel, and on actions taken to discipline safeguard measures, it would appear that the Uruguay round's grand bargain delivered what had been promised to the developing nations by the advanced economies. To complete the analysis, it is necessary to review the degree to which the objectives of the advanced economies were achieved.

NEW TRADE RULES

New rules pertaining to trade in services were a high priority in the Uruguay Round negotiations for the advanced economies, particularly the United States. U.S. service providers were innovative and globally competitive. By the late 1990s, the Joint Economic Committee of the U.S. Congress estimated that the U.S. surplus in services trade had grown to over U.S.$80 billion annually, and offset over 40% of the U.S. merchandise trade deficit.

In many ways, the new trade rules applying to services were soft and subject to ongoing negotiation, but they did firmly establish the authority of the new WTO in this important segment of world trade. The new rules did establish a basic MFN obligation to foreign service suppliers, subject to specific exceptions that must be reviewed after five years and terminated after ten years. Also established were provisions on market access and national treatment that were not general obligations, but specific commitments made in national schedules. Importantly, the rules required full transparency in the publication of all relevant laws and regulations.

A basis was established for progressive liberalization through successive rounds of negotiations and the development of national schedules. Particular attention was to be paid to financial services, telecommunications, and maritime commerce. A good example of this policy of progressive liberalization was the WTO agreement of 1998 regarding foreign ownership of national telecommunication carriers. The new rules also provided for conciliation and dispute-settlement services through the WTO Council on Services.

The Uruguay Round also provided the initial overview by the GATT/WTO of the process of foreign direct investment through new rules pertaining to trade-related investment measures, or TRIMS. As noted earlier, by the 1990s, global commerce conducted through the global affiliates of multinational enterprises has surpassed the volume of commerce conducted through international trade. However, prior to the Uruguay Round, foreign direct investment lacked the supervision that GATT and later the WTO provided to international trade.

The new Uruguay Round rules provided that no GATT/WTO member could apply any TRIM inconsistent with GATT Article III, which guarantees national treatment, or inconsistent with GATT Article II, which prohibits quantitative restrictions on imports. The new agreement also prohibits the establishment of minimum "local content" and/or "trade balancing" export requirements, which are either mandatory or voluntary to achieve eligibility for an investment incentive. These rules were particularly directed at the concept of "import substitution" schemes, which will be discussed later. The requirements contained in the new rules were mandated to be met by advanced economies in two years, and by the developing nations in five years.

As discussed in Chapter 3, the Uruguay Round also authored strong new regulations as regards the protection of intellectual property rights. The advanced economies are currently bound by these rules, and the developing nations will be bound by 2000.

Looking at the major elements of the grand bargain between the advanced economies and the developing nations on which the Uruguay Round strategy was based, it is fair to say that the objectives of each were addressed and accomplished to a considerable extent, even though additional work needs to be done and additional agreements need to be crafted.

The round was probably successful beyond the hopes of its most ardent supporters. Through its new trade rules and market access agreements, future world trade is predicted to be at least 25% greater than it otherwise would have been, and world income is predicted to rise by more than U.S.$500 billion annually. It should now be easier to efficiently convert global inputs into outputs for global markets and to create conspicuous customer solutions.

In addition to the objectives of the grand bargain, the Uruguay Round also attempted to tighten the curbs on the use of the antidumping laws and codify allowable subsidies and countervailing duties, as described in the chapter on trade distortions. But, without doubt, the round's most visible accomplishment was the creation of the new World Trade Organization.

THE WORLD TRADE ORGANIZATION

The new World Trade Organization was authorized in the GATT Uruguay Round and came into existence on January 1, 1995, essentially replacing and enhancing GATT. The WTO assumed authority for all agreements and arrangements concluded under the auspices of GATT and the complete results of the GATT Uruguay Round. The WTO maintains a permanent secretariat of about 425 persons in Geneva, Switzerland. Its first director general was Renato Ruggiero, the former trade minister of Italy. Interestingly, the candidate favored by the United States for this post was former President Carlos Salinas, of Mexico.

In 1999, at the scheduled end of Renato Ruggiero's term as director general, a lively dispute arose as regards the choice of his successor. The WTO membership appeared to be equally divided between Mike Moore, a former prime minister of New Zealand, and Supachai Panitchpakdi, deputy prime minister of Thailand. The stalemate was broken on September 1, 1999, when Mike Moore became director general for a term of three rather than four years, to be followed by Supachai Panitchpakdi, also with a three-year rather than four-year term.

The WTO differs from GATT in several important aspects. The

WTO is more global in membership. It includes all previous GATT members plus others who have joined subsequent to its formation, and still others are considering accession. Importantly, those seeking membership include China, a former GATT member, and Russia. The addition of these two large nations to the WTO would strengthen global trade disciplines.

The WTO enjoys a broader scope of responsibilities than did GATT, bringing trade in services, trade in agriculture, the protection of intellectual property rights, and trade-related investment measures into the system for the first time. The WTO is a full-fledged international organization, as opposed to GATT's provisional treaty status. It enjoys a similar status for international trade and investment that the IMF and the World Bank enjoy in global finance. Unlike the GATT, which conducted its trade negotiations in eight separate and distinct negotiating rounds, the WTO is a forum where nations can continuously negotiate the exchange of trade and investment concessions to further lower such barriers around the world.

THE WTO DISPUTE-SETTLEMENT UNDERSTANDING

For many years the United States and other world nations had been concerned about GATT's long, drawn-out dispute-resolution procedures, the softness of these GATT decisions, and the difficulty of enforcing even these decisions. It was this phase of the GATT process that caused many nations to believe that GATT no longer appropriately served the needs of international trade. The new dispute-settlement process contained in the WTO Dispute-Settlement Understanding could well be one of the Uruguay Round's most enduring accomplishments.

Under the WTO process, a dispute-settlement body (DSB) will exercise the WTO's authority on all trade agreements. An accused nation must enter into consultations with its complainant within thirty days after a trade complaint is lodged with the DSB. If no resolution is achieved in sixty days, the complainant may request a dispute-settlement panel consisting of three persons of appropriate backgrounds and experience from countries not party to the dispute. The dispute-settlement panel must complete its work in six months. The panel report is then sent to the DSB to adopt or not adopt in sixty days. The DSB cannot adopt it if one of the parties notifies its intention to appeal.

If the DSB panel report is appealed, an appellate body would be convened to review the issues of law covered in the panel report. These procedures would be limited to sixty days. The resulting report must be adopted by the DSB and accepted by the contesting parties within thirty days following its issuance. Within forty-five days, the accused party must advise the DSB its intention with respect to implementing the adopted recommendations. Failure to comply with the recommendations would place the party in violation of WTO rules and permit the aggrieved party to implement sanctions of its choice.

Even though this new WTO dispute-settlement process could extend to as long as sixteen months, it is still a step function improvement in timing compared to the previous GATT process, and most cases should not last the full sixteen months. In addition, the WTO decision, when finally realized, is readily enforceable, which has not always been the case of GATT decisions. To date, the WTO process seems to be working well and on a timely basis.

Even though the United States was one of the most vocal critics of the former GATT dispute-settlement system, and one of the principal architects of the strengthened and accelerated WTO system, the new WTO system ran into considerable opposition in the U.S. Congress when it was asked to ratify the agreement. Congressional opponents argued that participation in the WTO agreement would result in the loss of U.S. sovereignty on trade issues, would deny the ability of the United States to use, in the future, its own trade statutes, such as Section 301, and that the new trade dispute-settlement mechanism would deprive the ability of the United States to ignore the reports of international panels and appellate bodies, as it had done in the days of GATT.

Apparently, what these Congressional opponents sought was a trade system that bound U.S. trading partners but not the United States. This is not a likely international outcome based on the predominance of the United States in international trade and foreign direct investment. In such matters, if the United States does not go along, it does not happen. However, it is well to remember that it was also the U.S. Congress that refused to ratify the International Trade Organization immediately after the end of World War II, which resulted in the substitution of GATT in 1947.

Sovereign parties to any agreement voluntarily and provisionally relinquish some freedom of action to advance the common good. However, decisions of the WTO do not change U.S. laws, nor the right of the United States to apply its own laws. If the United States decides to invoke its own laws in violation of a WTO appellate body finding, the United States must then be prepared to accept retaliation, as authorized by the WTO to the aggrieved party.

To resolve this issue, the U.S. ratification of the Uruguay Round Treaty contains a provision, that if and when the United States loses two consecutive WTO dispute-settlement cases, a panel of U.S. judges will be convened to review the equity of the legal conduct of the cases, and to make recommendations regarding continued U.S. participation. Hopefully, such a need will never arise. A U.S. withdrawal from the system would destroy an organization for which the United States fought long and hard to create. Thus far, the United States has fared well in its WTO dispute-settlement cases. It has not won them all, but fortunately it has not lost two consecutive cases. The conclusion of the GATT Uruguay Round and the creation of the World Trade Organization is a milestone on the road to a more open, integrated global economy in which global inputs can be efficiently converted into outputs for global markets, and conspicuous customer solutions can be readily created.

THE WTO "BATTLE IN SEATTLE"

The WTO scheduled a membership meeting for early December 1999 in Seattle, Washington, to develop an agenda for the first trade negotiating round to be sponsored under its auspices, and the first such negotiating round to be convened since the conclusion of the GATT Uruguay Round in 1994. The Seattle site was specifically chosen because of that city's high level of international commerce, its high level of export-related employment, and the desire of the current U.S. administration to enhance a legacy of international trade expansion.

Unfortunately, for a host of reasons, the WTO meeting rapidly deteriorated into what was quickly labeled the Battle of Seattle, as aggressive protestors promoting numerous, narrow, self-interest agendas successfully engaged in the process of civil disobedience to essentially prevent the WTO meeting from happening and to provide further credence to the American reputation for lawlessness.

"The Battle of Seattle" evolved from a broad range of contributing causes, not the least of which was the apparent inadequate preparation by the city and its police force for protestor violence, which had been broadly predicted in advance, and by its inability to successfully deal with a semiprofessional anarchist group based in nearby Eugene, Oregon, which successfully infiltrated the individual protest groups and provided the catalyst for much of the violence and destruction of property.

The long and often acrimonious debate that preceded the selection of a new WTO director general to replace Renato Ruggiero had essentially divided the WTO membership into advanced nations, which supported Mike Moore of New Zealand, and developing nations, which supported Supachai Panitchpakdi of Thailand. Even though this issue was compromised in September, it provided Mike Moore with only three months to prepare for the Seattle meetings, following a six-month leadership vacuum at the WTO, initiated by the departure of Renato Ruggiero on April 1.

Prior to the Seattle meeting, inadequate consensus had been reached between the major trade players regarding an appropriate agenda for a new trade round. This was particularly true in the case of the United States, the European Union, and Japan. The nations of the developing world, in particular, were spooked by the U.S. administration's statements, made just prior to the session, that trade sanctions might be an appropriate measure to enforce international labor and environmental standards. The developing world, which regards these issues as thinly disguised instruments of advanced-nation protectionism, responded with a stiffened resolve against advanced-nation trade demands. And, finally, the United States apparently misread the united opposition of the developing world to the U.S. refusal to renegotiate its antidumping laws and found itself standing alone on this issue.

The Seattle debacle demonstrated the potential vulnerability of institutions such as the WTO to narrow-issue protectors intent on protecting their self-interest privileges at the expense of other world nations and their workers. In most cases, these protestors had failed to achieve their agendas through the democratic process. Hopefully, "the Battle in Seattle" will not cause the WTO to become the institution of choice for the exploitation of every activist agenda in the world.

Much of the Seattle rhetoric was crafted to misinform about the mission, structure, operations, and authority of the WTO. It portrayed the WTO as a covert, supranational body with the power to overrule the sovereign laws of individual nations and accountable only to itself. Unfortunately, all too often, such misinformation, repeated with sufficient frequency and unrefuted, can eventually become accepted as truth.

The WTO, like its predecessor, GATT, is not a supranational government body. It is a voluntary association of nations through which its members seek to improve broad global economic welfare by reducing barriers to international trade, liberalizing foreign direct investment, and protecting intellectual property rights. Similar to GATT, the WTO is a creature of its own members, no more and no less, and is directly accountable to its membership.

Decisions of the WTO do not overrule the sovereign laws of its member nations, nor the rights of these nations to apply their own laws. However, since sovereign parties to any agreement voluntarily and provisionally relinquish some freedom of action to advance the common good, nations that choose to invoke their national laws in opposition to WTO agreements must be prepared to accept sanctions, as authorized to an aggrieved party. The critics' complaint regarding covertness, no doubt, directly relates to the strengthened WTO Dispute-Settlement Understanding, which was one of the major improvements sought over previous GATT procedures. Since much of the dispute-settlement process is not fully transparent, the WTO should seriously consider making this process more open in order to address the concern of covertness.

More than fifty years of global trade expansion has confirmed what economists have long taught: that increased trade between nations creates prosperity and that prosperity promotes peace. The WTO should be about realizing the opportunities inherent in globalization, but at the same time it must be about helping to guard against its risks, in that economic integration cannot successfully advance if too many individuals and their interests are left behind.

In recent years, many of the world's international financial institutions, such as the World Bank, the IMF, and the Inter-American Development Bank, have broadened their criteria for lending programs to include labor issues and other concerns, such as combating corruption. These institutions now formally prohibit the use of forced labor, exploitive child labor, and corruption in the projects they support. This is a model that the WTO could well consider. These would not be the types of global labor and environmental standards so feared by the nations of the developing world, and which they see as a pact to deed to the advanced-nation politicians the right to dictate developing-country domestic policies in order to appease special interest groups in the advanced nations.

The basic thrust of the GATT Uruguay Round was to trade enhanced access to the markets of the advanced nations for the commodity and simple manufactured products of the developing world in exchange for developing-world acceptance of new rules for trade in services, trade-related investment measures, and the recognition of intellectual property rights. Having bargained for this enhanced access, the developing world is going to strongly oppose global standards, which it believes are intended to permit advanced-nation politicians to diminish this enhanced level of market access.

The process of economic development in the developing world has demonstrated its ability over time to improve labor and environmental standards in these nations. However, their interest in these subjects usually does not develop until they achieve a level of economic sufficiency that permits them to address the basic needs of their populations. Accelerating the economic-development process is certainly in everyone's best interest, both in the developing and advanced world, and will eventually make possible higher levels of global standards, as sought by many. But will this happen if advanced nations use arbitrary global standards to restrict the access of developing-world products to their markets and significantly reduce the ability of the 85% of potential global consumers, who live in the developing world, to purchase products and services from advanced nations?

Chapter 8

Foreign Direct Investment

At the very beginning of this book, it was emphasized that a critical issue in the creation of conspicuous customer solutions is the enlightened confluence of technology, innovation, trade, and foreign direct investment. Thus far, this book has discussed the subjects of technology and innovation and has rather extensively reviewed the salient facets of international trade. Beginning with this chapter, attention now turns to foreign direct investment, a complementary business tactic to international trade in assisting private-sector enterprises to optimize the global deployment of their resources.

As foreign direct investment has evolved, so have the titles assigned to those enterprises that engage in this practice. Initially known as multinationals, this term was deemed inappropriate by the United Nations in that it implied enterprises whose capital came from many nations and not just the fact that these enterprises conducted business and had affiliates in several nations. The United Nations felt the term *transnational* was more appropriate, and this term or its abbreviation, TNC, is still used in all U.N. publications on the subject.

In more recent years, the term *global enterprise* has become more popular to describe an organization that operates in multiple nation states through foreign direct investment affiliates that are sub-

ject to some degree of central control over the strategic aspects of
the business, such as production and pricing policies, choice of
technologies, appointment of key personnel, and the determination of markets to be served. Additionally, a global enterprise's
competitive position in one market must significantly impact, and
be impacted by, its competitive position in other nations, through
the creation of economies of scale, the enhanced ability to serve
other global enterprises, or the marketing leverage achieved
through transferable brand names.

Enterprises that operate in multiple nation states but whose
competitive position in one nation has little influence over its competitive position in another, are more properly defined as multi-
domestics. This is no way infers that multi-domestics are inferior
to global enterprises, it just recognizes the fact that multi-
domestics do not benefit from cross-border market leverages to
the extent enjoyed by global enterprises. In many ways the distinction is more academic than practical, in that the commercial,
political, and social issues faced by both are quite similar.

Foreign direct investment is a private-sector investment and can
be either the private flow of equity or loan capital. It is measured
as a flow of solely financial resources and includes reinvested
earnings as well as the cross-border flow of new funds. Very often,
foreign direct investment is packaged with the flow of other valuable resources, such as management skills, technical assistance, or
marketing know-how. Regardless of the great value of these other
resources, they are not included in the calculation of foreign direct
investment, which is limited solely to financial flows.

Foreign direct investment is differentiated from portfolio investment by the intent of the investor to exert some form of control over the acquired asset. The United States, Germany, and
Sweden presume an equity investment of 10% or more to indicate
the intention to exert some form of control. Australia uses a standard of 25%, and the United Kingdom and Japan use other criteria. Thus, the levels of foreign direct investment flows from
different nations are not necessarily directly comparable.

Because of its nature, a foreign direct investment is normally a
less liquid and longer-term investment commitment than a portfolio investment. If you were to purchase one hundred shares of
a mutual fund that specializes in foreign equities you could be in
and out of that position every day, every week, and every month

because of the small size of your investment and the liquidity of the mutual fund. However, when you purchase 10% or more of the equity of a specific foreign enterprise, your investment no longer enjoys the liquidity basis of that mutual fund and the capability for frequent sales and purchases. Furthermore, the purchase of a sizable portion of the equity of a foreign firm is usually done in support of the objectives of a reasonably long-term business plan, which should not be changing on a daily or weekly basis.

Foreign direct investment is only one component of global capital flows. Another is portfolio flows, which can consist of investments in government or corporate bonds, corporate equities, mutual funds, and other forms. Additionally, there are other short-term flows of trade credits, bank loans, financial leases, currency, and bank deposits.

Once a foreign direct investment has been made, it can, of course, change in value. This can be caused by the retention of earnings or losses or by the market value appreciation or depreciation of the asset. These two events can also apply to domestic investments. But, in the case of foreign direct investment, changes in the currency exchange rate between the currency of the investor and the currency of the country in which the asset is located can also impact the value of that investment in the accounts of the investor. Such change has no impact on the value of the asset in the currency of its home country, and there are complex rules that apply to the accounting treatment of this change in value in the accounts of the foreign investor. Tracking annual FDI flows is a relatively straightforward procedure. However, incorporating subsequent value changes is a much more difficult art and results in less accurate estimates of the current values of long-term, cumulative FDI stocks.

As noted in Chapter 1, in recent years, annual FDI outflows from advanced nations normally exceed FDI inflows into those nations, whereas FDI inflows into the developing world usually exceed FDI outflows from the developing world. According to the Institute for International Economics, during the 1990s, FDI from the advanced economies provided for more than 30% of the capital inflow into the twenty-nine major emerging nations.

Late in the 1990s, cross-border mega-mergers, such as British Petroleum/Amoco (U.S.$55 billion), U.S. West/Global Crossing

(U.S.$51 billion), Daimler/Chrysler (U.S.$40 billion), and Deutsche Bank/Banker's Trust (U.S.$10 billion), began to accelerate the growth and distort traditional FDI inflows and outflows. The United Nation's Commission on Trade and Development (UNCTAD) estimates that in 1999, global FDI flows rose to U.S.$800 billion, up 72% from 1997, and FDI inflows into the advanced economies rose to U.S.$640 billion, a 134% increase from 1997. At the same time, the level of FDI inflows into the developing world continued to decline to around U.S.$160 billion as an aftermath of the Asian financial crisis of 1997–98.

This cross-border mega-merger phenomenon has also raised the ratio of global FDI inflows and outflows to gross fixed capital formation from a traditional 5%–6% to 7%–8%, and the ratio of global FDI inward and outward stocks to GDP from a traditional 9%–10% to 11%–12%. Of interest is that the FDI annual inflow into Japan is virtually negligible, at U.S.$3 billion, and only 10% of the FDI outflow from Japan. Additionally, the annual FDI inflow into China is U.S.$45 billion, more than 30% of all FDI flowing into the developing world. In the case of China, this is not a Western-world phenomenon, but represents mainly capital inflows from ethnic Chinese living outside of China, primarily in other Asian nations.

According to UNCTAD, by 1998, foreign direct investment stock, that is, the cumulative value of all foreign direct investment in the world, stood at U.S.$4.1 trillion, with U.S.$2.8 trillion of this being in the advanced economies and U.S.$1.3 trillion in the nations of the developing world. Enterprises based in advanced economies were the source of U.S.$3.7 trillion of this investment, and enterprises based in the developing world were the source of U.S.$0.4 trillion. Again, it is interesting to note that 20% of all inward FDI stocks in the developing world are in China.

As noted above, in 1998, the global stock of FDI was estimated to be U.S.$4.1 trillion, consisting of parent-firm equity contributions, intracompany loans, and reinvested earnings. These FDI affiliates also contained over U.S.$10 trillion in other assets, consisting of funds such as bonds or loans raised either in the domestic capital market of the FDI host nation or in international capital markets. Financing also came from equity contributions by local partners or shareholders in the host nation.

In 1998, sales revenues of these FDI affiliates reached U.S.$11.4

trillion, which included a gross product (value added) of U.S.$2.7 trillion, or 24%. This low percentage of value added is due to many of the affiliates receiving significant levels of value added from other units of the same corporate family, plus the fact that in FDI affiliates engaged in lifting, gathering, and transporting natural resources, these activities are of small value compared to the value of the underlying resource. The fact that FDI affiliate assets exceed annual sales revenues is also due to natural resource projects, where front-end capital costs are often a multiple of annual sales revenues. Nevertheless, it is still notable that FDI sales revenues in 1998 exceeded world exports by a factor of 1.7/1, thus establishing international production as the primary source for reaching global markets.

GLOBAL INVESTMENT TRENDS

Again, according to UNCTAD, in 1998 there were approximately 60,000 global parent enterprises that had equity positions in approximately 508,000 FDI affiliates. About 50,000 of these global parents were headquartered in one of the advanced economies, and approximately 10,000 were headquartered in nations of the developing world. About 95,000 of the FDI affiliates were located in one of the advanced economies, and about 415,000 were located in nations of the developing world. Approximately 145,000, or over 40% of all FDI affiliates in nations of the developing world were located in China.

The largest sources of FDI outflows among the advanced economies have been the United States, the United Kingdom, Germany, France, Japan, and the Netherlands. The advanced economies with the largest FDI inflows have been the United States, the United Kingdom, France, and Belgium/Luxembourg. The advanced economies with the largest FDI outflow surpluses have been Germany, the United Kingdom, Japan, and the Netherlands.

The developing nations with the largest FDI inflows have been China, Brazil, Singapore, Indonesia, and Mexico. The developing nations with the largest FDI outflows have been Hong Kong, Singapore, South Korea, Taiwan, and Thailand. With the exception of Hong Kong and South Korea, all of the developing nations have had a net FDI capital inflow.

A very high percentage of all FDI flows and stocks are between, and in, what is known as the triad nations of the United States, the European Union, and Japan. In 1998, UNCTAD estimated that over 70% of FDI global outward stocks and 45% of global inward stocks were concentrated in these three economies. Confirming this dominance of FDI by the triad nations, eighty-nine of the one hundred largest global parent enterprises, based on total foreign assets, were headquartered in the triad, with forty-five being in the European Union, twenty-seven in the United States, and seventeen in Japan. The headquarters of the remaining eleven were distributed between Switzerland (5), Canada (3), Australia (1), South Korea (1), and Venezuela (1).

THE PROLIFERATION OF INTERNATIONAL TRANSACTIONS

Obviously, the driving force behind the rapid growth of foreign direct investment has been the sharp increase in the level of intense global competition, the desire to more efficiently convert global inputs into outputs for global markets, and to facilitate the creation of conspicuous customer solutions. For many years, enterprises have exported goods and services to foreign markets. These same enterprises now invest abroad to better serve these same markets and to obtain more favorable access to the factors of production.

The more broadly global enterprises organize their international operations, the more frequently they engage in a wide variety of international transactions. The best known of these are probably international trade and foreign direct investment. International trade is best described as a cross-border, third-party transaction between an unaffiliated buyer and seller. A foreign direct investment is a cross-border investment in which the investor seeks some form of control over the acquired asset in order to better serve a foreign market or obtain more favorable access to the factors of production.

Another potential type of international transaction is two-way, cross-border, intrafirm trade, in which FDI affiliates belonging to the same global parent export to and import from each other and the global parent. It is estimated that today over one-third of all international trade is between parent and affiliate or affiliate and affiliate. In the case of U.S. global parents and their FDI affiliates, the international commercial integration is even greater, and approaches 50% of all U.S. cross-border merchandise trade being conducted on an intrafirm basis.

International cooperative agreements and strategic alliances have also become popular as a means of gaining access to technology, manufacturing know-how, and marketing expertise. So have nonequity involvements such as technology licensing, franchising, and management contracts. Technology licensing was fully discussed in Chapter 3. Many service industries use franchising agreements to develop international markets, such as those for fast food and car rental. Many of the international hotel chains combine the management acumen, brand-name recognition, and the global reservation resources of the chain to operate hotel structures owned by local nationals in foreign countries. And finally, the technique of international subcontracting can be employed to access more efficient sources of supply at lower levels of asset investment.

Historically, international trade has represented the principal method of accessing foreign markets. As noted earlier, in 1998 the sales revenues of foreign direct investment affiliates were 1.7 times the level of global exports. Exports include, of course, trade to unaffiliated buyers as well as intrafirm trade between units of the same global enterprise. The sales revenues of FDI affiliates include value added in that unit plus value added received from other units of the same enterprise in the process of intrafirm trade. As the degree of commercial integration continues to expand between affiliates of the same global enterprises, it is going to become more and more difficult to differentiate between the contributions of international trade and foreign direct investment to global economic growth. Integrating trade and FDI within an enterprise's business plan has become a very successful corporate strategy and is likely to continue. No one can argue with its positive results, but the ability to continue to separate the statistics of trade from those of FDI is going to continue to decline and become an effort of questionable value.

THE DOMESTIC IMPACT OF FDI

In the last chapters of this book, the impact of foreign direct investment on the nations of the developing world is reviewed. But what is the impact of foreign direct investment in the advanced economies? Is the impact positive or negative? It is probably both, but is the net balance positive or negative?

What are the nationalities of the principal foreign direct investors in the United States? The biggest investment group are enterprises from the United Kingdom, without doubt, a reflection of the earlier colonial affiliation. The second largest FDI stock in the United States is owned by Japanese enterprises, which just recently surpassed investors from the Netherlands, who are now number three.

Why does foreign direct investment come to the United States? Probably, most important is that the United States is viewed as a safe place to invest funds. Property rights are well respected, the nation is militarily strong, and has stable political and legal structures. In many aspects, safety is more important to portfolio investors than FDI investors, in that FDI investors are usually searching for a specific asset required to support their business plan.

The American marketplace is a big marketplace. There are significant advantages to having production facilitates close to current and prospective customers. A presence in the American market provides improved intelligence regarding changing technologies and changing U.S. customer needs and desires. Once inside the American market, an enterprise is protected against the erection of access barriers to the U.S. market.

Finally, the sum of U.S. expenditures exceeds U.S. domestic output. The demand for foreign credit and equity in the United States reflects the excess of U.S. investment over savings. U.S. rates of return appear superior to those in markets where the biggest proportions of investable savings have been accruing, that is, Germany and Japan.

In more recent years, the U.S. media and the U.S. government have become increasingly concerned with the massive flows of foreign capital into the United States. In reacting to such concern, it is well to remember that these inflows reflect the endemic U.S.

current account deficits. This concern has, unfortunately, been concentrated on the inflow of foreign direct investment, which is more visible, rather than on portfolio investment. Foreign direct investment, by its very nature, is a less liquid, longer-term investment than portfolio and other short-term investment, and therefore less subject to sudden flight. Portfolio and other short-term investment is much less permanent and much more volatile. This particular lesson was learned the hard way by many of the Asian nations during the Asian financial crisis of 1997–98. A strong case can be made that if a nation's current account deficit requires that nation to import significant sums of foreign capital, its financial stability will be aided by having a major portion of that capital come in the form of foreign direct investment rather than portfolio and other short-term investment.

Other alleged hazards to the United States from high levels of inward foreign direct investment are the "dumbing down" of U.S. affiliates, the threats that these affiliates pose for U.S. national security, and the concern that the U.S.-based affiliates will not act responsibly. Fears that these U.S. affiliates will be "dumbed down" and that the good corporate jobs will be kept in the home country of the investor are not supported by the economics of the U.S. market. The U.S. labor market is an expensive market, and there is not a surplus of skilled, low-wage workers. U.S. subsidiaries of foreign firms typically pay comparable wages to their domestic counterparts and do comparable levels of research and development. The United States is not a good location in which to seek to perform low-wage, busy-finger work.

As regards the threat of foreign direct investment affiliates in the United States to U.S. national security, the United States has restrictions on FDI in certain business sectors that are sensitive to its national security. These restrictions are contained in the Atomic Energy Act, the Federal Aviation Act, the Federal Communications Act, and others. Under the Exon-Floria provision of the Trade and Competitiveness Act of 1988, the U.S. president was authorized to prohibit or suspend investments and/or acquisitions that in his judgment threatened U.S. national security. One of the few times this authority was invoked was during the administration of President George Bush, when a Japanese ceramics firm was seeking to acquire a similar U.S. firm that had a subsidiary engaged in nuclear work. The issue was resolved with the U.S. enterprise divesting its nuclear subsidiary, and the acquisition of the remaining ceramic assets was then approved.

The question as to whether U.S.-based foreign direct investment affiliates will act responsibly is best answered by considering what factors encourage responsible behavior. In general, responsible behavior is mandated by market forces, civic forces, the media, and government rules and regulations. It is hard to imagine that a foreign investor would purposely negate the value of his FDI affiliate by engaging in practices that would alienate any of these constituencies, which are so important to the success of his investment. As a matter of fact, most FDI affiliates are so conscious of their "foreignness" that they are much more concerned about responsible behavior than many of their domestic counterparts.

What are the positives of foreign investment into the United States? To begin, foreign capital inflows into the United States have financed U.S. investment in excess of U.S. savings. This capital inflow is the counterpart of excess U.S. imports over exports. If the United States had to rely on its savings alone, higher interest rates, prices, inflation, and reduced consumption would result. More generally, international capital flows permit a more efficient allocation of world resources and allow savers and investors to globally diversify their assets and liabilities. Well-placed foreign investment can help a country augment its trade advantages by building large, more productive facilities to increase exports or decrease imports. And through the establishment of new firms, jobs can be created in the economy.

There are also broad socioeconomic benefits to foreign investment into the United States that bring valuable resources into the U.S. economy, such as technology, management, and marketing expertise. Such investment encourages domestic entrepreneurship by the purchasing from or subcontracting to local suppliers, the training of new workers in new skills, and the spreading of this knowledge to others. Labor benefits by obtaining more jobs and higher wages, consumers benefit from lower prices and a wider choice of quality products, and the government benefits from higher tax revenues.

One question frequently asked is whether the foreign ownership of real estate is a domestic negative. Many countries erect barriers to such foreign purchases of lands in specific areas. This issue arises occasionally, such as when investors from the Middle East were purchasing U.S. real estate, U.S. citizens were buying property in Canada and other countries, Germans were buying property in Switzerland, and most recently when Japanese investors were buying trophy properties in the United States, such as the Pebble Beach Golf Course, hotels in Hawaii, and Rockefeller Center in New York City.

As these most recent Japanese purchases turned out, it was the investors who suffered the negatives, when after a very short time these properties had to be sold at distress prices. The Japanese Ministry of International Trade and Investment (MITI) now cautions Japanese real estate investors against buying trophy U.S. properties, the purchase of which causes distress among the U.S. population. Such advice would have been more valuable to the potential Japanese investors if they had received it prior to making their purchases.

Foreign ownership of U.S. real estate today stands at approximately 5% of the total. Is this bad for the United States? To some extent it depends on which 5% is foreign owned. At the moment it does not seem to be a problem, and if you are a U.S. owner of real estate you would probably like to have foreign bidders available to bid on any properties that you might elect to sell.

To the extent that inward foreign direct investment provides the United States with needed capital and foreign exchange, trade-related investment benefits, and broader socioeconomic benefits, it can be a major plus to the economy of the United States. The more it focuses on productive investment opportunities, the greater the net benefit likely to ensue. Most assuredly beneficial is foreign direct investment that funds the production of exportable products and/or import substitutes, the foreign exchange earnings or savings from which exceed the servicing cost of the foreign investment.

Chapter 9

FDI: Who Needs It?— Almost Everyone

In the past quarter century, the pace of change in the nature and composition of international trade has accelerated. International trade no longer conforms to the textbook model of an exchange of British cloth for Portuguese wine. The specialization of national firms in particular products is being increasingly replaced by the globalization of production in which different processes for the production of individual goods and services are performed in different countries. Said another way, global inputs are increasingly being sought to be efficiently converted into outputs for global markets. An automobile sold in America may have been designed in Japan, assembled in Canada or Mexico, and include parts manufactured in Taiwan, Brazil, or just about anywhere.

Globalization, therefore, means increased trade in parts, components, and semifinished goods. It also implies an increase in intrafirm trade as single global enterprises move components and semifinished goods from their facilities in one country to those in another. In short, the traditional horizontal pattern of trade in final products is being overtaken by the form of vertical trade in which countries specialize in different parts or stages of the production process for individual products. It is for this reason that trade and investment are no longer alternative means of penetrating foreign markets. They are now mutually supportive tactics in a strategy of optimizing the global deployment of an enterprise's resources.

The globalization of production is made possible by two developments: the rapid advances in transportation and communications technology (including data processing), which have enabled managers to coordinate widely dispersed activities and the steady reduction of trade barriers under the aegis of GATT, which has made possible the movement of components and semifinished goods across national frontiers with a minimum of penalties in the form of tariffs or other restrictions.

Another major development in international trade is the rapid rise in the importance of trade in services. In addition to travel and transportation, it includes a wide range of other services, such as advertising, accounting, financial, insurance, architectural, construction, engineering, educational, medical, and many others. Many types of service exports cannot be provided effectively through cross-border exports, but only through the establishment of local presence in the foreign countries in which service is provided. In the service sector, trade policy and foreign direct investment policy converge. For the United States, trade in services is extremely important. U.S. service providers are very innovative and intensely globally competitive. The Joint Economic Committee of the U.S. Congress estimates that since the early 1990s the annual export of U.S. services has more than doubled, to U.S.$250 billion, and the U.S.$80 billion annual surplus in services trade has offset more than 40% of the U.S. merchandise trade deficit.

International trade and foreign direct investment are inextricably linked. They can no longer be regarded simply as alternative means by which a producer seeks to penetrate a foreign market. As a result of the globalization of production, trade and foreign direct investment have increasingly become integral elements in a firm's unified strategy for optimizing the deployment of its resources worldwide. The symbiotic relationship between trade and investment is reflected in the increasing proportion of foreign trade that takes place within individual global firms. Today, it is believed that well over 50% of all U.S. merchandise exports consists of intrafirm trade. Of this total, 60% consists of U.S. exports by American firms to their foreign affiliates, and 40% consists of exports by foreign-owned firms in the United States to their parent companies abroad or to other affiliates.

Traditionally, trade has been considered the "engine of growth" in the world economy, and a comprehensive framework was established within the GATT to reflect that perspective. Today, with much of foreign trade occurring as intrafirm transactions and with technology flows heavily associated with foreign direct investment, FDI is increasingly a driving force in international economic growth. Worldwide flows of foreign direct investment have risen at unprecedented rates in recent years. During the 1990s, the cumulative stock of global FDI has grown more than twice as fast as the global exports of goods and services, which, in turn, have grown more than twice as fast as world economic output.

In evaluating the role of foreign direct investment, it is important to remember that at the macroeconomic level, net international flows of capital are simply the counterpart of national imbalances in trade in goods and services (in essence, the current account balance). A trade-surplus nation is a net capital exporter; a trade-deficit country is a net capital importer.

Most international movements of capital consist of financial portfolio and other short-term flows rather than direct investment in the sense of the acquisition of a controlling interest in a new or existing foreign enterprise. A host nation can encourage foreign investment through incentives such as tax concessions, and it can discourage it through disincentives such as mandating majority local ownership. Additionally, the imposition of severe restrictions on imports, or arbitrary rules of origin, can impel a foreign exporter to establish a local production facility to serve a protected market. Japanese auto plants in the United States are an example of this, as are the EU semiconductor rules of origin, which have induced semiconductor investment in the EU.

The recent trends in capital flows have coincided with shifts in the location of world current-account imbalances. The emergence of the United States as the major host country and Japan as a major home country for foreign investment has paralleled their roles as the world's principal trade-deficit and trade-surplus countries, respectively.

At the microeconomic level, the link between trade and FDI is obvious. If an exporter establishes a plant abroad, he may no longer export that product from his home country, or the foreign affiliate may export components back to the corporate parent, who then assembles and re-exports the product.

During President Ronald Reagan's administration, the White House issued a policy statement to the effect that "foreign investment flows which respond to private market forces will lead to more efficient international production and thereby benefit both home and host countries." Unfortunately, that sentiment was not shared by all, either in the United States or abroad. Questions have constantly been raised about the effects of foreign direct investment on the economies of both home and host countries.

Steep increases in foreign direct investment inflows have raised fears in the United States to "the buying up of America" by foreigners, and the loss of control by the United States of its own economic destiny. Similar cries were heard about FDI in the United States from the Arab states during the two oil shocks in the 1970s, and for many years from Canada and many developing countries about U.S. foreign direct investment in their countries.

Concerns about foreign direct investment are not limited to inward flows. Fears exist that FDI abroad causes the loss of domestic production, jobs, and exports from the domestic economy. Specifically, in manufacturing, it is often assumed that in the absence of FDI abroad, production would remain within the domestic economy. In most cases this is an unrealistic assumption in that much investment abroad is intended to serve the markets of the host country in which the investment is made.

The basics of international capital flows are that the net inflow of capital from abroad in all forms is simply the counterpart of a nation's deficit in its current account. The inflow can be reduced only by narrowing the current account deficit. An essential condition for reducing the external deficit is the narrowing of the gap between domestic income and the sum of domestic public and private expenditures.

As long as a gap persists, a domestic economy will draw on capital from abroad. Therefore, the question is not whether a country wants or does not want foreign investment, but rather the form in which the inflow of capital is to take place. Is the investment to be in liquid financial instruments or in more permanent and less volatile FDI assets?

In recent years, there has been a marked trend toward the liberalization of policies regarding foreign direct investment in both the advanced economies and developing nations. Despite large structural imbalances in the international accounts of many of the advanced economies, controls on FDI have not been used as a major means of correction. The main reliance has been on macroeconomic adjustments (fiscal, monetary, and foreign-exchange policies). However, trade measures have also been used, and these have had secondary effects on FDI flows.

In the developing world, confrontation has been replaced by pragmatism and the active encouragement of foreign direct investment. In most countries, the former attitude of conflict has changed to an understanding of the mutual benefits inherent in cooperation between host governments and foreign affiliates.

Today, foreign direct investment from the advanced economies is virtually free of home-government controls. The main concern of home governments regarding outward FDI is to achieve fair treatment for their home-based global enterprises in foreign markets. From the point of view of global enterprises, there are many legitimate reasons for making investments in other countries. Among these are the ability to bypass foreign import barriers (either actual or threatened), avoid high transportation costs, adapt products more easily to local tastes and standards, achieve greater efficiency in the integration of production, marketing, and after-sales service, establish a "presence" in a foreign market, or achieve a more cost-effective production base.

In earlier decades, particularly the 1970s, many of the advanced economies restricted outward FDI through foreign-exchange controls designed to address balance-of-payment issues. The worldwide liberalization of capital markets began in the 1970s, as Canada, Germany, and Switzerland abolished all restrictions on foreign capital movements in 1973, followed by the United States in 1974, the United Kingdom in 1979, Japan in 1980, France and Italy in 1990, and Spain and Portugal in 1992.

Many of the more economically advanced developing countries also began opening up their capital markets in the early 1990s, partly under pressure from the United States and the IMF, but also in recognition of the potential benefits that more open markets could confer. What few remaining controls exist today deal with extraordinary political or economic circumstances. Not only have global FDI outflow restrictions been relaxed, now virtually all advanced economies actively promote outward FDI, especially into nations of the developing world.

A good example of such promotion is the U.S. Overseas Private

Investment Corporation (OPIC), which is an agency of the U.S. government. OPIC organizes investment missions for enterprises that are at least 50% owned by U.S. nationals. It offers project loans or loan guarantees for FDI into the developing world. It also insures FDI against noncommercial risks (expropriation, currency controls, etc.) and provides a wide variety of investor services designed to reduce the risks associated with FDI.

Japan, with its high rate of domestic savings and gigantic current-account surpluses, is probably the most active advanced economy in promoting outward FDI by its enterprises. In total, eight Japanese agencies sponsor promotional programs for outward FDI. Created in 1950 to finance Japan's exports and provide credits for raw material imports into Japan, the Export-Import Bank of Japan is today the most active institution in FDI promotion, with over 40% of its current activities devoted to financing outward FDI loans. From 1988–92, outward FDI from Japan was fifteen times the FDI inward flow into Japan. In 1996, the U.S. inward and outward FDI flows were essentially equal.

Inward FDI, however, has long been subject to control, and some countries have retained their review requirements to protect national, political, and economic objectives from the impact of global firms. Although the criteria and procedures for screening have largely been liberalized, some domestic sectors (transport, shipping, and broadcasting) remain closed or maximum level of investment limited.

The surge of Japanese foreign direct investment has been less influenced by policies of the Japanese government than by threats of foreign protectionism and the strong yen. The situation on foreign direct investment into Japan is less clear. MITI has created an office to promote inward FDI and has arranged favorable financial accommodations. However, as noted earlier, in the five-year period from 1988 to 1992, Japan's outward FDI was fifteen times inward FDI into Japan.

In the developing world, recent changes in laws and regulations confirm the liberalization of inward FDI. However, conflicting attitudes toward foreign enterprise is reflected in a mixture of measures to either encourage it or control it. The technological gap between developing and developed countries has widened, and FDI is viewed as a vehicle to narrow the gap.

Currently, the burden of foreign debt is consuming local investment capital in many developing nations, especially Asian nations. With new commercial lending practically nonexistent, FDI plays a vital role in relieving this capital shortage. For many years, FDI from the advanced economies has supplied two-thirds of the total private resource flows to the developing world.

In many ways the developing world faces a dilemma. The recent debt crisis has dealt a blow to the conventional wisdom that commercial borrowing had the advantage of providing unencumbered resources in contrast to foreign control over local enterprises in direct investment. Developing nations have regretfully discovered that borrowing results in more, not less, external intervention. When debt-repayment schedules could not be met, countries had to submit to severe austerity programs as a condition of debt-rescheduling and IMF emergency financing. The alleged problem of foreign control over this or that enterprise receded in the wake of the far-reaching economic, social, and political consequences of externally negotiated conditionality.

Despite the liberalization of developing-nation views toward FDI, these countries have not abandoned the principle that entry and operation of global affiliates should be subject to some control to ensure compatibility with national objectives. Thus, institutional agreements for registering and authorizing FDI have generally remained in effect, but with administrative simplifications and improvements. Performance requirements still abound, but the willingness to accept international arbitration is growing.

The threat of expropriation was a major deterrent to FDI in the developing world prior to the 1980s. Following World War II, emerging from colonialism and without training for dealing with foreign enterprises, the developing world was highly suspicious of foreign enterprises. These fears were apparently confirmed by the highly publicized role of ITT in allegedly destabilizing the Allende regime in Chile during the early 1970s.

However atypical that episode, it intensified the developing world's hostility toward foreign enterprise and gave legitimacy to widespread acts of expropriation, many as public demonstrations of political sovereignty. From 1970 until 1975 there was an average of fifty-six expropriations per year. A peak of eighty-three was reached in 1975. As an indication of the improvement in this situation, since 1980 the average has been three per year.

The declining importance of the nationalization issue also reflects other view changes in the developing world as well as changes in attitudes toward foreign investors. State ownership of enterprises is declining in Asia and Latin America, in contrast to growth in earlier periods. Additionally, insurance against expropriation risks can be bought today from international investment guarantee agencies. The Multilateral Investment Guarantee Agency (MIGA) of the World Bank, which began operations in 1988, is unique in that it is sponsored by the major capital exporting countries plus fifty-five developing nations.

Few developing nations have paid much attention to outward FDI policies in that they were subsumed under general capital control policies that were quite restrictive. These nations needed more capital, not less, and their economic development was already constrained by the lack of foreign exchange. In such nations capital flight has been seen as a major problem that has led to controls, which often leak. In more recent years, nations such as Taiwan, Korea, Singapore, and Hong Kong have achieved balance-of-payments surpluses, and these regulatory controls have been relaxed to permit outward FDI by their local enterprises. However, the Asian financial crisis of 1997–98 brought an end to much of this, and some Asian nations were again reapplying capital controls.

It is important to remember the two basic facts that underlie foreign direct investment in the developing world. First, all countries have the right to limit the establishment of global enterprises in their countries and to regulate the operation of these enterprises. Second, global enterprises can't be forced to invest in a specific country; they must be attracted. To the extent that these two extreme positions can be mutually accommodated, general global prosperity and the rate of world economic growth will benefit.

FDI's LACK OF INTERNATIONAL DISCIPLINE

Today, foreign direct investment lacks the international disciplines that GATT and now the World Trade Organization afford trade, and that the World Bank and the IMF afford finance. As the importance of service trade increases, the right of establishment and national treatment become critical factors to the ability to establish national presence.

In the mid-1970s, the developing world launched a campaign from within the United Nations for a "new international economic order" (NIEO) inspired, in part, by the spectacular success of the Organization of Petroleum Exporting Countries (OPEC). The objective of NIEO was to transfer wealth from the advanced economies to those of the developing world, much as the OPEC cartel had done with oil revenues.

The central theme of NIEO was "to tame the multinationals." At the same time, academic and labor critics appeared in the United States and Europe. For trade unions, the enhanced mobility of global enterprises significantly reduced their bargaining power and increased the prospects of "runaway plant" job loss. In this atmosphere, in the mid-1970s, both the OECD and the United Nations launched efforts to negotiate international agreements on foreign direct investment. In both organizations, the focus was at least as much on restraining the conduct of the global enterprises as ensuring their fair treatment by local governments. However, it soon became apparent that there was not, at that time, a viable international consensus favoring guidelines for foreign direct investment.

Finally, in the Uruguay Round of the GATT, the United States sought a broad GATT proscription on trade-related investment measures, or TRIMS, which were described in Chapter 8. Following the conclusion of the Uruguay Round, it appeared that the newly authorized WTO could be a convenient forum to develop a comprehensive international investment accord. The likely elements of such an accord could be the right of establishment (subject to exceptions for reasons of public order and national security), the right of national treatment, the right to expropriation compensation, an open transfer of funds, and access to dispute-settlement facilities such as the Permanent Court of Arbitration in the Hague, the International Chamber of Commerce Court of Arbitration, or the World Bank's International Center for the Settlement of Investment Disputes.

All of these elements have appeared in one form or another in bilateral and multilateral agreements to which many of the advanced economies and some of the developing nations are already signatories. By the end of 1997 there were over 1,300 such investments treaties involving over 160 nations. In late 1996, Renato Ruggiero, the then energetic director general of the WTO, began promoting the idea of a WTO-sponsored negotiation for international investment rules. Ruggiero argued that the proliferation of bilateral and regional treaties threatened the coherence and predictability of global trade and investment.

According to Ruggiero, a WTO investment accord would benefit the nations of the developing world in several ways. It would guarantee a secure and stable environment for investors. It would curb the beggar-thy-neighbor investment incentive schemes, which favor nations with deep pockets. It would offer protection from discrimination by regional pacts, and it would take into account the interests of the developing nations, especially the poorest, which receive little investment.

Late in 1995, the member nations of the OECD had undertaken to draft global rules for the treatment of international investment, on the basis that OECD member nations are the major sources for such investment. The product of the OECD deliberations, which surfaced in 1998, and which was known as the Multilateral Agreement on Investment (MAI), was bitterly attacked by environmentalists and trade unions, which claimed that the accord would permit global enterprises to ride roughshod over environmental rules and reduce labor standards. It was also attacked as a product solely of the advanced economies with no input from the nations of the developing world. In April 1998, the twenty-nine OECD members agreed to a pause in these negotiations to permit a period of assessment and further consultation. In December 1998, the MAI was put to a merciful death. Mr. David Henderson, the former chief economist for the OECD, ascribes the MAI failure to two main interrelated sources of concern: fierce disagreements amongst OECD members and the rising anxieties of multiple nongovernmental organizations (NGOs).

As noted earlier, there is an inexorable link between international trade and foreign direct investment. These are not alternative means of achieving global competitiveness, they are mutually supportive tactics in an overall strategy of optimizing the global deployment of an enterprise's resources.

Today, foreign direct investment lacks the international disciplines that GATT and now the World Trade Organization afford trade, and that the IMF and the World Bank afford finance. As

the importance of service trade increases, the right of establishment and national treatment become critical factors in the ability
to establish national presence. The time is ripe for a new foreign
direct investment accord under the auspices of either the newly
formed World Trade Organization or the OECD. The broader con-
stituency of the WTO probably makes it the more natural organization to undertake this task, and the WTO intends to include
the FDI issue in its year 2000 international trade and investment
negotiation package.

Chapter 10

FDI and Developing Nations: Risks and Jurisdictions

Thus far, we have discussed how the enlightened confluence of technology, innovation, trade, and investment favor the efficient conversion of global inputs into outputs for global markets and the creation of conspicuous customer solutions. In these last three chapters we will now look at this process from a slightly different perspective and review how the process of foreign direct investment also assists in the expansion of economic development in the nations of the developing world. The larger the size of the global pie, the larger the slices available for those who elect to compete.

The principal sources of foreign direct investment today are global enterprises that were earlier described as enterprises that operate in multiple nation states, that have some degree of central control over the strategic business aspects, and whose competitive position in one nation significantly impacts, and is impacted by, their competitive position in other nations.

According to the May 1998 IMF World Economic Outlook and the UNCTAD 1997 World Investment Report, the developing world consists of 156 nations, of which 128 are considered developing nations and 28 considered countries in transition. These nations account for almost 85% of the global population and 45% of global GDP on a purchasing power parity basis. Their exports of goods and services account for 23% of global trade, and they are home to 66% of all foreign direct investment affiliates. On a market exchange rate basis, their GDP per capita averages only 5% of the advanced economies, although this is not a meaningful figure in view of the vast differences between the individual nations of the developing world.

The twenty-nine advanced economies are productive and economically similar. Their GDP per capita range, on a PPP basis, is about 3/1, from U.S.$13,000 to U.S.$34,000. The average is U.S.$25,000, and the U.S. GDP per capita is U.S.$29,000. The nations of the developing world are much less homogeneous and include the former COMECOM nations, the high GDP per capita member nations of OPEC, the NICs of Southeast Asia, and the abject poor nations, such as Somalia, Chad, and Ethiopia. Even when excluding the member nations of OPEC, the range in GDP per capita between the nations of the developing world, on a market exchange rate basis, is almost 100/1. And herein lies an important paradigm for global enterprises operating in the developing world. Because of the vastly different economic and other circumstances existing in these nations, a single business concept is not universally applicable throughout the developing world. One size does not fit all.

In spite of these vast internal differences, the developing world is an emerging commercial opportunity, and foreign direct investment is a key ingredient to realizing that opportunity's full potential. As noted previously, 85% of the world's population resides in the developing world, and its share of the world's population continues to grow. The GDP of the developing world has been growing twice as fast as the GDP of the advanced economies. Compared to the advanced economies, the needs of the developing world are vastly underserved. Although most developing countries have been steadfastly embracing the principles of liberalization and market force economics, many still remain with vestiges of central planning and control. As the 1997–98 Asian financial crises revealed, much of the "Asian Miracle" was based on economic artificialities that favored producers over consumers. Global enterprises not prepared to deal with such impediments are best advised not to apply.

Chapter 10 will address the developing-nation issues of the risk inherent in foreign direct investment and the means to minimize that risk. It will also deal with the subject of legal jurisdiction and dispute settlement. And finally, it will examine the critical relationship between human-resource development and the process of foreign direct investment.

INVESTMENT RISK

If you were to ask an international businessman, experienced in foreign direct investment in the developing world, his view of the three greatest risks inherent in FDI, there is a good possibility that his response would be: first, instability; second, instability; and third, instability. Being more specific, he would probably say: political, social, and economic instability, as well as complex, drawn-out, and arbitrary bureaucratic procedures. To which he might add: unpredictable and abrupt changes in the conditions of operation, or in ownership or remittance regulations after the investment costs have been sunk. The potential exposure to asset expropriation might also be mentioned, but not as likely today as twenty years ago.

In addressing the instability issue, there is no better test than reviewing the political, social, and economic history of the host nation under consideration and of talking to current domestic and foreign direct investors in that nation. Of particular importance is the level of domestic investment. If the domestic investors, who understand that nation far better than you do, are currently sending their funds to numbered accounts in Swiss banks, the message cannot be very encouraging regarding a foreign direct investment.

It is also possible to purchase risk assessments for specific nations from risk-rating firms. One such firm is International Business Communications Ltd., of London, England, which combines in its risk assessment: political risk, which it describes as the extent of government corruption and the extent to which economic expectations diverge from reality; financial risk, which reflects the likelihood of losses from exchange controls and loan defaults; and the economic risks of inflation and debt service charges. Not surprisingly, this annual rating places most of the advanced economies in the very-low- and low-risk categories, and most of the developing nations in the moderate, high and very high categories. However, there are a surprising large number of developing nations in the low- and moderate-risk categories.

Such ratings obviously describe the risk as the analysts currently perceive it and make no provision for changes that might occur after the investor's funds are sunk. In developing nations, changes in conditions are inevitable, as these countries proceed through the social and economic transformations that constitute the process of economic development. But social and economic change is not unique to developing nations. Global enterprises face social and economic change in their own home markets and have learned to adapt. To succeed in the developing world, these same enterprises must also learn how to adapt to legitimate responses to altered conditions in the nations of the developing world.

As earlier discussed, the U.S. Overseas Private Investment Corporation provides insurance coverage for most noncommercial risks, including expropriation, in nations of the developing world. It insures only U.S. investors and does not insure commercial risks. Errors of management are still for their own account. Additionally, the World Bank sponsors the Multilateral Investment Guarantee Agency, which covers many of the same risks as the OPIC, and which is available to all enterprises whose home governments are World Bank members. MIGA is funded by both advanced nations and fifty-five nations of the developing world.

In the final analysis, the investment decision is going to rely on the expected rate of return compared to the inherent risk, including the exposure of front-end sunk costs. However, in foreign direct investment, the investor does not always have the choice of going to the option of least apparent risk. Sometimes the assets he seeks only exist in nations of relatively high risk and he must seek investment strategies to minimize his exposure to such risks over the term of the investment. Sometimes it is just necessary to select the least undesirable choice out of a bad lot of options and purchase the best investment protections available.

Complex, drawn-out, and arbitrary bureaucratic procedures in a developing nation are a sign that although that nation is proceeding toward a market-based culture, it is still enmeshed in a system of central planning and control. Such complex procedures obviously delay and add risk to investment decisions. They also create a favorable climate for the nurturing of corruption and bribery. Normally the more numerous, arbitrary, and detailed the regulations, the greater the potential for bribery. Although many of these bribe opportunities may be categorized today as facilitating payments, or "grease payments," the opportunities for real bribes also exist. For enterprises based in the United States, these numerous, complex, and arbitrary bureaucratic procedures have long represented an additional investment risk because of the enterprise's exposure to the dictates of the U.S. Foreign Corrupt Practices Act. Hopefully, after 1998 all enterprises from OECD-member nations will be constrained by like measures.

In assessing investment risk, a review of the specific developing nation's risk history can be instructive. Also, comparative OPIC and MIGA insurance rates are good indicators. Insurance against expropriation is not expensive, and relative coverage rates are a good measure of investment risk. However, in the final analysis, the behavior of local investors is most important. If these people, who know the domestic investment climate far better than you do, are investing their funds elsewhere, it should give you some cause for concern.

LEGAL JURISDICTION

Negotiating an agreement to invest in a specific developing nation is no different than any other kind of negotiation. Each party has certain specific rights. Each developing country has the right to limit the establishment of global enterprises in their country and to regulate their operation within the limitations of bilateral investment treaties and rules imposed on members of the WTO by the TRIMS agreement. However, global enterprises can't be forced to invest in a specific country, they must be attracted. Somewhere within these specific rights, the parties should be able to negotiate the details of their agreement, with the eventual terms being determined by the leverages and options enjoyed by each.

Once the agreement has been negotiated, whose laws will apply to the operations of that FDI affiliate? Invariably, it will be the laws of the host country. Many nations of the developing world subscribe to the "Calvo Doctrine," named for the Argentine foreign minister who created it. The Calvo Doctrine simply states that foreign affiliates, once established in a foreign country, give up diplomatic access to their parent's home country and must settle all matters within the laws of the host developing nation. In years past, when expropriation was rampant, many home governments of global enterprises attempted to intercede in such disputes, but with limited success. Since local laws will govern FDI affiliates, the investor should know the laws before he invests, and also be sure he receives the national treatment, which is mandated beginning January 1, 2000 for all nations that are WTO members.

But it is important to remember that in the developing world, the process of economic development is going to continually alter social, economic, and political circumstances. Thus, the initial investment agreement should specify the manner in which changes to that agreement are to be accommodated in order to adjust to changed conditions in the developing country. And, if at all possible, that investment agreement should include a dispute-resolution mechanism with guaranteed access to third-party arbitration.

There are, today, a large number of international facilities for the settlement of investment disputes. Among these are: the World Bank's International Center for Settlement of Investment Disputes (ICSID), the International Chamber of Commerce's Court of Arbitration, the Permanent Court of Arbitration in the Hague, and the WTO, as regards trade-related investment measures. In addition, there are regional arbitration facilities, such as the Arbitration Center in Kuala Lumpur, the Arbitration Center in Cairo, and various Asian/African legal consultative committees. Traditionally, the developing world was fearful of such institutions, which they regarded as having been created by the advanced economies to promote the values of the advanced economies. However, that image is changing, and agreement to access these bodies can now be regularly negotiated.

To summarize the subjects of legal jurisdiction and dispute settlement regarding FDI into the developing world, the FDI affiliate will most likely be governed by the local laws of the host country. The TRIMS agreement should provide exemption from such trade-distorting requirements as minimum local value-added contents and minimum export requirements. The investment accord should be specific as to whose laws apply to the interpretation of the accord and how future investment accord changes will be accommodated. And the investment accord should also establish a specific process for dispute resolution with guaranteed access to specific third-party arbitration panels.

HUMAN-RESOURCE DEVELOPMENT

When considering FDI, it is well to remember that a major reason that a developing nation seeks FDI is the beneficial impact that foreign investment can have on enhancing the skills of its workforce and preparing that workforce for more complex and higher-paying jobs. This desired benefit applies not only to factory operators but also to office workers, technicians, managers, and executives. The developing nations want their citizens to be able to earn access to all job levels in these FDI affiliates, including the very top one. This developing-world objective need not be in direct conflict with the interests of the global investor. In fact, it could be very much in his interest also.

In reality, the vast majority of even the largest global enterprises still have a culture deeply rooted in their country of origin. Changing that monolithic culture is one of the biggest challenges in becoming a genuinely global enterprise. Since most expatriate managers tend to merely extrapolate that domestic culture, most global enterprises now strive to reduce their reliance on this management source. Most global enterprises intend not to have more expats, but to have fewer of them.

In some respects, the expatriate is a holdover from the old days of the multinationals, when managers were sent out from the head office, much like colonial governors, to run the overseas possessions. The current aim is now to employ local managers who have been taught the culture of the global enterprise. What is needed is a balance by which the strength of local knowledge is combined with global reach. Said another way, you must think global but act local.

Developing nations expect that improved job skills resulting from FDI will also impact skills in other domestic enterprises linked to those foreign affiliates as suppliers and/or customers. Domestic competitors' job skills are also expected to benefit from the increased competitive pressure provided by the new FDI affiliate. In view of these skill-enhancement objectives, it is not unusual for developing nations to offer, as an investment incentive, free training of its local labor using the new enterprise's own parts and processes. Such skill enhancement is often regarded as being as valuable as the technology and capital assets also transferred in the process of foreign investment.

To initially staff new FDI affiliates in the developing world, local nationals are often recruited and sent to the global enterprise's home plant or to the plant of one of its other FDI affiliates to learn the enterprise's products, processes, procedures, and policies. N.V. Phillips at one time operated a developing-world demonstration factory in Eindhoven, the Netherlands, where skills were taught on labor-intensive processes that had been designed to produce products in the developing world using a maximum of developing-nation labor and other inputs. The people, thus trained in this demonstration facility, were then sent back to their home country, where they trained other local factory and office workers, often under a host-government-sponsored training program.

Well-conceived and well-executed work-training programs can enhance the FDI affiliate's subsequent financial performance and hasten its achievement of profitable operation. However, such programs can also favorably influence the attractiveness of an investment proposal to a host government and make it possible to more easily achieve certain other objectives in the investment agreement sought by the investor. Quite often, in the case of FDI in the developing world, there are major language barriers. However, in today's competitive world, a global enterprise must be prepared to manage locally in the local language—another good argument for limiting the use of expatriate staff.

In spite of the act-local objective, quite often in the process of FDI into the developing world, initially senior management and technicians must be parachuted in from either the home of the global enterprise parent or from one of the enterprise's other foreign affiliates. Initial selection and planning for the ultimate replacement of these expatriates is a very important part of the human-resource plan. Some globals are fortunate to be able to recruit these top people in-country, which is one advantage of a joint venture with local nationals who are experienced business people, speak the local language, know the local environment, know other good local managers, and can be readily taught the enterprise's business.

Initially, new FDI affiliates are well received by host governments because of the many potential benefits they bring to the economy. There is initially a honeymoon period. However, after time, such matters as a large cadre of expatriate managers enjoying high pay and benefits, or the realization that the FDI affiliate is enjoying large tax concessions begin to raise equity questions with the local population and to cause friction. Because of this, many developing nations attempt to establish limits on the number of expatriates to be imported and their maximum length of stay. Obviously, the developing nations want these jobs available to their citizens as quickly as possible.

To the extent that such expatriate limitations retard the development of the FDI affiliate, this is not in the best interests of either the investor or the local government, which wants that level of benefits resulting from a successful local venture. What the developing nation must understand is that fewer expatriates for shorter periods of time also benefits the foreign investor, as long as there are not too few expatriates for too short a period to achieve success. Expatriates are a very high cost for a global enterprise. Not only is their compensation high compared to in-country scales, but it must be augmented by foreign housing allowances, sometimes a car and driver, servants, perhaps private schools for their children who don't speak the local language, annual home leave, and other expenses. The quicker the expats can do their training tasks and be withdrawn, the better it is for the global enterprise. Thus, as regards the expatriate situation, the interests of the global enterprise and the developing nation coincide, and this needs to be explained and well understood at the very beginning of the investment relationship if problems are to be avoided.

Unfortunately, when expatriate employees are withdrawn and replaced by local nationals, another culture shock occurs when the compensation of the local replacements is based on local practice and quite often only a small fraction of the total compensation paid to the expatriate he is replacing. This is another situation that needs to be discussed and understood by all parties at the outset of the investment accord.

The expatriate is being paid to create a new business in a culture foreign to him and his family and to train the people necessary to run that new business after his departure. His skill level is demonstratively higher than that of his replacement, whose task is to manage a business created by his predecessor with a staff trained by his predecessor. Also, as a local national, the replacement does not need many of the benefits enjoyed by his predecessor, which were associated with that predecessor's and his family's foreignness to the local environment. Both the enterprise and the local government benefit from the skills of this expat. But the higher his skills, the greater his compensation. Hopefully, his higher skills can reduce the time he needs to remain and improve the skills he can impart to his successor. Properly explained at the outset, a minimum expat staff can bring great benefits to the start-up of an FDI affiliate and impart the skills necessary to the local staff, who will then manage the affiliate.

In thinking of management paradigms for the staffing of FDI affiliates, it is obvious that the initial expat chief executive must have skills well beyond just the internal functioning of the affiliate. He must understand public affairs, have reasonable political skills, and be able to deal with an unfamiliar environment. He should also be reasonably fluent in the local language.

A well-documented and well-executed employee training program will not only produce enhanced business operations but will provide the local affiliate with improved leverage in negotiating the initial investment accord. The local government must be made to understand the initial need for expatriate employees, and that the global enterprise is just as interested as the developing nation in minimizing the length of their stay because of their high cost. A limit on expats and their length of stay can also be counterproductive. Their early departure can retard the commercial development of the FDI affiliate and prevent the achievement of many objectives the developing nation sought from this initiative.

Finally, right up-front, the local staff and local government must be made to understand the high compensation and foreign allowances paid to these expats. They must also understand that local replacements will be paid according to local scales. However, these local replacements should be made aware that they are eligible for more senior positions elsewhere in the global enterprise, including positions in its parent organization.

Chapter 11

FDI and Developing Nations: Technology and Finance

This chapter continues the review of how foreign direct investment not only contributes to conditions that improve the opportunities to efficiently convert global inputs into outputs for global markets, but also contributes to the economic development of developing nations, which, in turn, assists in expanding global market demand. Specifically, this chapter will consider the subjects of technology transfer and investment financing. Included in this review will be the methods and terms of technology transfer, the appropriateness of the technology transferred, and the technological activities undertaken in the nations of the developing world. As regards finance, the review will include an analysis of the balance of payments effects of inward FDI, the sources of FDI funds, and the contentious subject of transfer pricing between a global enterprise and its FDI affiliates.

In today's world, the capabilities of technology, innovation, and job skills are generally regarded as the key resources to the competitive strength of both individual enterprises and individual nations. It is appropriate to recall the earlier quoted comment of Harvard Business School Professor Michael Porter, that "national prosperity is created, not inherited. It does not grow out of a country's natural endowments, its labor pool, its interest rates nor its currency values, as classical economics insists. A nation's competitiveness depends on the capacity of its individuals and enterprises to innovate and upgrade."

As the global environment becomes more competitive and rapid technological changes result in shorter product life cycles, an enterprise's ability to generate new or improved products and find new processes that reduce production costs becomes an increasingly important determinant of the enterprise's competitiveness.

A large portion of the research-and-development expenses that form the basis for technological progress in today's global economy is concentrated within the corporate research staffs of global enterprises. The United Nations estimates that 75% to 80% of all global civilian R&D comes from this source, and on the basis of patents granted, the world's seven hundred largest industrial firms, most of which operate globally, account for half of all of the world's commercial inventions.

Therefore, it should be no surprise that inward FDI and non-equity modes of participation by global enterprises in the developing world are important and powerful methods of technology transfer and technological capability building that benefit the economic performance of these nations. It is for this reason that developing nations often regard technology transfers and skill enhancement as the primary benefits of inward foreign direct investment. The financial capital generated, mobilized, transmitted, and invested by global enterprises is, naturally, also an important contribution to a developing nation's output and productivity growth, but it is certainly not the only favorable factor in the process of foreign direct investment.

In most developing nations, the ratio of inward FDI flows to gross fixed capital formation seldom exceeds 10%, although it does reach 20% to 25% in China. Despite this seemingly small contribution to domestic capital stock, FDI capital is especially important for developing nations, on account of the diminished flow of other official and private capital in recent years from sources such as official development assistance (ODA) and commercial bank loans. According to the IMF, in the 1990s, private funds' flow into the developing world accounted for more than 3% of GDP, compared to 0.5% of GDP in official funds' flow. The Institute for International Economics notes that from 1992 to 1998, FDI provided 40% of all private capital funds to the twenty-nine major emerging markets. Now there is also within the developing

world a growing recognition of the need to balance loan and equity capital in private foreign capital inflows. This was certainly
a message from the 1982 and 1997–98 debt crises.

TECHNOLOGY TRANSFER

The absorption and application of technology in its broadest
sense, that is, the knowledge of how to perform useful activities
and how to make useful things, is at the heart of the economic
development process. Such valuable knowledge is not always protected by intellectual property rights, and is available in the public
domain. However, such knowledge needs a convenient vehicle to
effectively transfer it to a developing nation that seeks its acquisition. In many cases, the vehicle could be the process of foreign
direct investment, or, alternatively, through the services of a qual-
ified consultant. Later in this chapter, the relative merits of seeking
"unbundled" assets will be discussed.

In Chapter 3, the historical attitude of the developing world as
regards new knowledge, and the intellectual property right
thereto, was discussed. But regardless of the developing world's
former attitude regarding new knowledge and the protection of
intellectual property rights thereto, the grand bargain of the Uruguay Round of GATT provided to the developing world many
trade benefits that it sought in return for its acceptance of strengthened intellectual property rights, new rules regarding trade in
services, and trade-related investment measures. Developing
nations had until January 1, 2000 to comply with these new IPR
rules, and the very least developed nations have until January 1,
2006.

Technology can be transferred through a nonequity transaction,
such as licensing, or an equity transaction, such as inward foreign
direct investment. Licensing was initially popular in the developing world, when foreign commercial bank lending was more
readily available, as a means of acquiring technological assets
while at the same time avoiding the dominance of a specific economic sector by a foreign enterprise. This process of acquiring
"unbundled" assets had great political appeal to the governments
of many developing nations. But as these nations were soon to

learn, the "bundling" of "unbundled" assets also brought problems and liabilities.

The rationale for the purchase of "unbundled" assets was that it was a means to avoid potential foreign control over certain local economic sectors resulting from the process of foreign direct investment in those sectors. The 1982 and 1997–98 developing-world debt crises not only severely reduced commercial bank lending to the developing world, but also exposed these nations to drastic external intervention in their fiscal and monetary policies in order to qualify for IMF emergency funding. In many cases, this external intervention in the domestic economy exceeded the level of external intervention that would have been caused by foreign direct investment in one or more domestic economic sectors.

The developing nations also learned that the successful integration of "unbundled" assets often required more sophisticated national infrastructures and skill levels than then existed in many of these nations. Japan was often cited as a nation that had very successfully bundled "unbundled" assets. However, few developing nations could match the sophistication of the Japanese economic infrastructure or the Japanese skill level. Additionally, it is the rare global enterprise that devotes as much time and support to a licensee as it devotes to a licensee in which it also enjoys a significant equity interest.

Foreign licensing for U.S. enterprises is made more difficult by the ex-territorial application of U.S. antitrust laws, which prohibit U.S. licensors from including geographic limitations in licensing agreements with nonaffiliated companies. These nonaffiliated licensees seldom export back into the home market of their licensor, but could well export into the home countries or third-country markets of other foreign affiliates, or the licensees of the global enterprise. Since most developing nations pressure their domestic enterprises to export for balance of payment purposes, the issue is often exacerbated.

One remedy often used is to seek agreement from the licensee to use the international sales force of its global enterprise licensor as its marketing channel for sales outside of its own nation. Since most developing country licensees don't have sales channels to markets outside of their own country, such an agreement could be of substantial benefit to that licensee. If such a practice were to be challenged by the U.S. Justice Department, it would be necessary to be able to show that the licensee benefited from this arrangement and that it was not just a method to circumvent the intent of U.S. licensing law.

In equity transactions, such as foreign direct investment, technology can be transferred by means of the products selected, the processes used, or the capital equipment involved. To the extent that the capital of the FDI affiliates is used to purchase production process or capital equipment, it is easy to see the tight relationship between foreign direct investment and technology transfer. In equity transactions, technology transfer can be financed in several different ways.

A global enterprise can engage in a nonequity transaction with an enterprise in the developing world and license technology for a royalty on sales revenues or make an outright sale of the rights to that technology for a specific lump-sum fee. As just noted, this was the solution earlier preferred by the developing world. In an FDI affiliate that is only partially owned, a global enterprise can assess a technical service fee to the affiliate to compensate for the technology transferred in addition to the capital invested. This technical service fee could be a single lump-sum payment, a percentage of the affiliate's sales revenues, or a combination of the two.

Also in the case of a partially owned FDI affiliate, the global enterprise could contribute the desired technology as a credit against its equity investment. In the case of a wholly owned FDI affiliate, a technical service fee can be assessed for the technology transferred, or the technology value could be recovered through the profits from the affiliate. There are tax and perception issues to this method, which will be discussed under transfer pricing.

As noted earlier, the developing world has never been enthusiastic about paying for technology, even when the payment is from a local FDI affiliate to its parent global enterprise in another nation. The size of the technical service fee permitted or of the lump-sum purchase price permitted will depend very much on just how badly the developing nation wants that technology, or just how badly the global enterprise wants access to the market of the developing nation, and what alternatives are available to each. Very often, the most difficult negotiations regarding technology concern not its costs, but the restrictive conditions that may be attached to its use.

Typical restrictions attached to the use of technology are global or territorial limitations on exports embodying the technology. The United States prohibits U.S. enterprises from invoking such limitations. Other technology agreements require the purchase of all production inputs, equipment, and processes from the technology provider. Unless such limitations are required to ensure quality in valuable brand-name products, this would seem to be a particularly shortsighted policy. In the case of a wholly-owned FDI affiliate, it would also have the appearance of just moving profits between individual nations.

Finally, developing nations object to the grant-back provisions, which give the licensor all rights to the technology improvements achieved by the licensee. Grant-back privileges are not just contained in technology agreements with developing world enterprises. They exist in all licensing agreements, including those between enterprises when both are based in the advanced economies. Grant-back is a long-established custom of the technical licensing culture and not just a practice recently created to annoy developing-world governments.

Developing nations are frequently concerned about the appropriateness of the technology being transferred to the needs of the country. Quite frequently, developing nations demand only the newest and most sophisticated technologies to be transferred, when in many cases earlier and less expensive technology generations may best serve their needs and be more compatible with the in-country support resources. On the other hand, developing countries worry about technologies that produce too sophisticated and too elegant products that cater largely to the elite and do not meet the needs of most of the people in poor countries. This latter issue could be easily resolved by the assessment of a luxury tax on such products, to be applied against all producers of those products, be they local or an affiliate of a global enterprise.

Lastly, developing countries are concerned if the technological processes are compatible with the stage of the nation's economic development. Is the technology too capital-intensive in a nation of abundant cheap labor? The industrial composition of a nation's output is a major determinant of the capital-labor relationship. The production of clothing and textiles is labor intensive. Extruding synthetic fibers, a new technology, is capital intensive and can't be made labor intensive. Many global enterprises frequently adopt consumer products to local tastes, climate, and materials. This is just what the N.V. Phillips demonstration facility was designed to do. However, can the small sales revenues available from many developing-nation markets economically justify major product- and process-modification costs for each nation? Outside of the consumer field, product modifications for developing nations are minor. Pharmaceuticals require high and uniform standards. In autos and machinery, accessories may vary, but the basics are standardized to facilitate global parts distribution and servicing. Low-cost labor can be low quality. Capital-intensive processes may be necessary to ensure brand-name quality.

Overwhelmingly, global enterprises conduct the research-and-development activities that produce new ideas and new technology in their own home markets or in the home markets of their foreign affiliates that are located in the one of the advanced economies. The need to achieve economies of scale, the availability of large pools of scientifically and technically trained staff and better access to local universities, and science and technology centers encourages the concentration of these R&D activities in the advanced economies. The desire to co-locate a sufficient number of innovative personnel to achieve the cross-fertilization of ideas as well as intellectual stimulation also favors fewer locations of larger size.

Nevertheless, research-and-development activities are beginning to spread to certain countries in transition and to developing nations in which significant scientific capabilities already exist, such as India and Russia, where mathematics and software-writing skills are strong. But in most developing nations, high levels of research-and-development activities on the part of FDI affiliates are not common. Normally, the development of product and process technologies for these affiliates is performed for them by their global parents in one of the advanced economies.

However, there are opportunities for FDI affiliates to perform product-modification tasks and product-line maintenance functions, including the development of product "midlife kicker" enhancements. This is particularly true when an FDI affiliate's sales of a particular product represents a significant percentage of that product's global sale and the staff of the FDI affiliate is technically competent and innovative. This is often the case when a global parent wishes to replace the current product with a more advanced solution and assigns the global product-line management-and-maintenance responsibilities for the existing product to a foreign affiliate, which still has significant sales revenues of the product. Products do not expire all at the same time globally. While the global parent is busily introducing a new solution, one of its FDI affiliates can enjoy a profitable revenue stream in supporting the remaining base of the expiring product and selling into those markets where the product is still a satisfactory solution.

In the case of products that require installation and continuing field-service capabilities, in a given region of the world, it is also possible to assign responsibility for these functions to a specific FDI affiliate and to train that affiliate's staff to perform those functions. These activities create service exports for the developing nation where the FDI affiliate is located, and which are just as welcome to that nation as merchandise exports. Additionally, these are higher-skill exports than normally involved in manufacturing jobs, which create developing-nation merchandise exports.

Even though, for valid reasons, most new ideas and technology are developed in facilities in the advanced economies, there are many talented technicians, engineers, and scientists in the nations of the developing world. These valuable resources should never be considered captives of the FDI affiliates, but should enjoy access to higher-level technical tasks and training in units of the enterprise. A man once said that the best way to judge the true globalization of an enterprise is to count the number of different languages spoken in the parent's global headquarters and in its research-and-development laboratories. Foreign affiliates are a great farm system for the growth and development of future corporate leaders, and a valuable resource is lost if there is not upward mobility available to those who originally sign on with a global enterprise in one of its foreign affiliates.

Technology transfer is potentially one of the greatest benefits to be achieved through the process of foreign direct investment into nations of the developing world. It is a process of delivering bundled assets, as opposed to the developing world's rather unsuccessful earlier attempts to achieve the same result through the combining of unbundled assets. Through greater experience, the developing world is finding ways to deal with some of the perceived problems of technology acquisition. At the same time, global enterprises are finding ways to accommodate many of the technology concerns that were most bothersome to developing nations. There is sufficient mutual benefit to all parties from this process of technology transfer, which is well worth the coordinated efforts of all parties concerned.

FINANCE

Even though inward foreign direct investment into the developing world typically accounts for less than 10% of the developing world's gross fixed capital formation, it can have a significant impact on a developing nation. If a new FDI affiliate produces a product that was formerly imported and also exports a portion of its output, that FDI affiliate can positively influence that developing nation's current account balance. And given the developing world's paucity of foreign exchange, it is only natural that these nations have concerns about FDI's balance-of-payment effects.

In foreign direct investment, there is an initial inflow of funds in the form of equity capital and loans either from the foreign investor or from a foreign bank. The level of foreign loan inflow can be reduced by the substitution of domestic borrowing, but this is not always welcomed by developing nations. Following this initial inflow, there is then an annual outflow, which could consist of management fees, technical service fees, interest on foreign loans, and the importation of parts. Assuming the venture is successful, there should also soon be an annual repatriation of some portion of after-tax profits. Depending on the debt/equity ratio of the affiliate, its profitability, and its level of local value added, this annual outflow could be 20% or more of the initial capital-funds inflow.

All too often, the critics of foreign direct investment concentrate their attention on the annual fee, interest, and profit outflows and parts supply, and tend to ignore the initial investment inflow and the positive current account effects. The result that this produces is the claim that they like to make: that foreign direct investment takes more out of a country than it puts in. Additionally, these critics fault the global enterprises for borrowing locally, charging that their better credit ratings encourage local banks to lend scarce local capital to them and not to potential domestic borrowers.

With little doubt, most global enterprises can borrow money from international sources at considerably more attractive rates than they could borrow those same funds from banks in a developing country. However, borrowing in the local currency provides some hedge against adverse currency-exchange-rate movements at a cost of the after-tax differential between international interest rates and the rates applying in that developing country. If developing countries find this practice offensive, they can always set limits on local borrowing, either at a specific amount or as a percentage of the total capital being contributed from abroad. On the opposite side of the coin, many foreign direct investment critics claim that a global enterprise's excessive use of foreign borrowing, as compared to equity capital contribution, is really an attempt by the foreign enterprise to avoid local profit taxes or through high intracompany interest rates, management and technical fees to accomplish this same purpose. This leads directly into the final subject of this chapter: the issue of transfer pricing.

TRANSFER PRICING

Transfer pricing is, no doubt, one of the most contentious issues of foreign direct investment into the developing world, and refers to the pricing of goods and services in transactions between related firms, such as a global enterprise and its FDI affiliates, which do not benefit from the disciplines of an arms-length transaction between unrelated entities. According to FDI critics, transfer prices are manipulated to adjust intracorporate component prices to shift profits from high-income-tax nations to low-income-tax nations. Intracorporate prices on components are reduced when imported into countries with high import duties, and prices are raised on intracorporate component exports into countries with rigid profit controls. And finally, these critics claim that prices are raised on intracorporate imports into countries that restrict the size of technology royalties, prohibit corporate management fees, or limit the levels of profit repatriation.

Unfortunately, most finance majors in the world's most prestigious business schools are probably being taught that the above financial practices are appropriate when faced with the situations described above. In actual fact, the latitude for such mischievous conduct is significantly less than most critics would have you believe, due to the increasingly intense scrutiny of cross-border, intracompany transactions. Earlier it was mentioned that, today, 50% of all merchandise trade both from and to the United States is conducted on an intrafirm basis. This obviously creates pressure on the U.S. Internal Revenue Service (IRS), the U.S. Customs Service, and their counterparts in other world nations to ensure that each government is receiving its fair share of income tax and customs revenue.

Most global enterprises follow the philosophy of transferring products, parts, components, and services toward the market at minimum allowable transfer costs to preserve maximum pricing flexibility to the final member of the corporate family that sells the finished product into the marketplace. The minimum transfer cost acceptable to the IRS or its counterparts is probably somewhere between standard cost and standard cost plus 15%, to compensate for manufacturing variances and the overabsorbing burden effect of such transfers. These intracompany transfer costs are subject to review by the IRS and its counterparts, and by the customs agents in the market of destination.

Obviously, higher component prices would please the domestic tax authorities in the nation of their manufacture by shifting profit to that country. Likewise, it would also enrich the customs duties collected in the nation of destination, all at the expense of the exporting enterprise. Without doubt, there could be an incentive to manipulate transfer prices to offset certain constraints previously mentioned. But the local taxing authorities and the local customs agents don't permit multiple transfer prices for the same part going to different countries or different customers. Thus, the global enterprise is usually stuck with just one transfer price, which hopefully optimizes its tax and customs charge exposure, and at the same time achieves an attractive sales price for the final product when sold into a competitive marketplace.

A number of years ago, the OECD issued a very soft guideline for the conduct of its member nations in transfer pricing. The OECD said that "enterprises should refrain from making use of the particular facilities available to them, such as transfer pricing, that do not conform to an arm's-length standard, or modifying in ways contrary to national laws, the tax base on which members of this group are assessed." The OECD could very properly have added that developing countries should refrain from imposing excessive and inequitable conditions on the operations of foreign direct investment affiliates that encourage such practices. Fortunately, the one export price for each component or part has largely constrained these types of games.

There can be no doubt that foreign direct investment can be a plus for developing nations, particularly when the investment's impact on a nation's current-account balance, through reduced imports and/or increased exports, exceeds the annual servicing costs of that investment, including all cross-border fees, interest payments, and the repatriation of earnings. The additional economic development thus achieved enlarges the size of the global market and enhances the ability to efficiently convert global inputs into outputs for global markets.

Chapter 12

FDI and Developing Nations: Linkages and Patterns of Operation

As earlier discussed, the process of foreign direct investment into a developing nation has the capacity to transfer a bundle of important resources, including financial assets, technology, job-skill enhancement, management acumen, and improved access to world markets. In this chapter, the issues to be addressed are: incentives to encourage inward FDI, linkages between FDI affiliates and the local economy, the potential patterns for FDI affiliate operations, and the earlier economic development strategy of import substitution.

INVESTMENT INCENTIVES

To access the above described bundle of resources, many developing nations offer incentives for FDI into their nations. Frequently, the rationale that developing nations use for such incentives is the opportunity to capture the "wedge" between the return captured by the investor and the total economic return from the process of foreign direct investment. This total return, the share captured by the investor and the remaining wedge, are illustrated in Figure 12.1. Many economists would describe this wedge as the positive externalities of foreign direct investment,

Figure 12.1
The Total Return From FDI

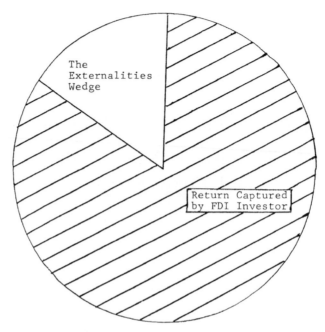

The
Externalities
Wedge

Return Captured
by FDI Investor

Source: United Nations, *World Investment Report*, 1997

usually realized in the form of social and economic benefits to the host nation.

Often included in these externalities are host-country benefits, such as additional domestic jobs of higher skill and wage levels, a wider variety of lower-cost, higher-quality goods and services available on the domestic market, higher local government tax revenues, improved foreign exchange balances, the introduction of new technology into the local market, the training of domestic workers in new, more valuable skills, the stimulation of local entrepreneurship through forward and backward linkages to the local economy, improved access to global markets through the international sales force of the investing enterprise, and increasing the competitive pressure on other domestic enterprises to improve their performance. These "externality benefits" from inward FDI provide the rationale for developing-nations' incentives to attract FDI.

Unfortunately, often these investment incentives derive not so much from the desire to capture the "externalities wedge" between the private and total return, but are offered to compensate for investment disincentives inherent in the economy of the developing nation, such as an overvalued exchange rate, an inadequate national economic infrastructure, or low education and job-skill levels. In this case, the offering of investment incentives is obviously a second-best solution, with fixing the disincentive being first best. In many cases, developing nations view foreign direct investment as an easier and less painful means of fixing the disincentives.

Finally, investment incentives can arise from fierce competition between nations to become the investment site for businesses that appear to be the leaders in innovation, new ideas, and sales revenue growth. However, so as not to suggest that developing nations are the only nations that get caught up in competitive investment-incentive contests, it is interesting to review the history of the advanced economies on this same issue.

A very specific case in point is the competition between nations and individual states in the United States regarding investment in automobile manufacturing facilities, as illustrated in Figure 12.2. This table not only demonstrates the pervasiveness of the investment-incentive practice in the world today, but also the escalation, over time, in the size of incentives paid per potential new job created. Also illustrated is the significant leverage enjoyed by potential host governments that have deep pockets.

Incentives for foreign direct investment can take many forms, including fiscal, financial, and others. Among the most prevalent are tax holidays during the early years of the investment, the authorization of rapid investment amortization, investment tax credits and outright grants, loans at below market interest rates, exemptions from import duties and duty drawbacks, subsidies for the training of local labor, and the financing of infrastructure, such as roads, housing, and power, water, and sewer facilities. Prior to the Uruguay Round, tariff protection for import substitution projects was frequently provided, as were subsidies for export. However, developing-country export subsidies are now banned under the World Trade Organization Agreement, as are quantitative import restrictions to protect new FDI investments. This will be further discussed in the review of the earlier strategy of import substitution.

Figure 12.2
Investment Incentives Offered for New Automobile Manufacturing
Plants

Year	Location	Incentive Source	Incentive Recipient	Incentive in Millions	New Jobs Created	Incentive Per Job
1983	Symrna	Tennessee	Nissan Motors	$ 33.0	1,300	$25,384
1984	Flat Rock	Michigan	Mazda Motors	48.5	3,500	13,857
1985	Spring Hill	Tennessee	GM	80.0	3,000	26,667
1985	Georgetown	Kentucky	Toyota	149.0	3,000	49,000
1985	Bloogmington	Illinois	Diamond Star	83.0	2,900	28,742
1986	Lafayette	Indiana	Fuju-Isuzu	86.0	1,700	50,588
1991	Setubal	Portugal	Auto Europa	483.5	1,900	254,451
1993	Tuscaloosa	Alabama	Mercedes-Benz	250.0	1,500	166,667
1994	Northeast England	U.K.	Samsung	89.0	3,000	29,675
1994	Spartanburg	South Carolina	BMW	130.0	1,200	108,333
1995	Castle Bromwich	U.K.	Jaguar	128.7	1,000	128,700
1995	Hambach	France	Mercedes-Benz	111.0	1,950	56,923

Source: United Nations, World Investment Report, 1995

Most evidence would indicate that the costs of FDI incentives offered by both developing nations and the advanced economies, under the pressure of intense internation competition, far exceed the dimension of the externalities wedge between the total and private return from foreign direct investment. In addition, there is strong evidence that incentives are not major factors in the selection of geographic locations for foreign direct investment. Much more important are factors, such as political, social, and economic stability, market size and growth rate, relative production costs, the levels of education and job skills, national economic infrastructures, and the regulatory framework.

Investment incentives would be an obvious issue for consideration in an international investment accord. But investment incentives would also be a tough issue for an international accord. In the previously illustrated case of investment incentives offered to gain automobile plants, most of the incentive providers were individual U.S. states and not sovereign nations. Could an international accord signed by the United States effectively bind the actions of its individual states in these incentive contests? However, a system of greater transparency of incentive offers, costs, and benefits would certainly be helpful. A requirement to report all such FDI incentive deals to the WTO for subsequent broad publication to all WTO members would shed greater light on this practice and the relationship between incentive benefits and costs.

What is the appropriate response for a global enterprise to the offer of investment incentives from a developing nation? Should that enterprise just wade into the trough of incentives with all four feet and sop up everything it can get? If not, why not? Some appropriate questions would seem to be: Why is the incentive being offered, and why might I not invest in that nation were it not for the incentive being offered? The next appropriate question would be to determine if the incentive or incentives offered are up-front, or do they have a long tail? Up-front incentives are infrastructure grants, subsidies for worker training, and outright investment grants. Long-tail incentives are tax holidays, investment tax credits, rapid investment amortization, loans at below market interest rates, and import duty exemptions.

One of the advantages of up-front incentives is that they are up-front. In most cases they are received well before the investor's costs are sunk and during the investment's honeymoon period, when all parties are enthusiastic about the project. These are incentives that are virtually impossible to cancel. Long-tail incentives continue well into the life of the investment, usually beyond the end of the honeymoon period and into the period when critics of the investment begin to raise issues about whether the foreign investor is paying his fair share of domestic costs, particularly when he is continuing to enjoy a holiday from income and real estate taxes. Long-tail incentives can be adjusted or canceled, especially if there is a change in government.

Depending upon the nature of the investment, the relative benefits of up-front and long-tail incentives may vary, and the long-tail variety may be much more valuable to the investor. However, to the extent possible, there are many advantages to maximize the use of the up-front incentives. Most foreign direct investors regard investment incentives as a nice frosting on an otherwise sound investment opportunity. But they don't consider that this frosting compensates for an otherwise rancid investment opportunity.

ECONOMIC LINKAGES

Foreign direct investment linkages are the forward and backward linkages between the FDI affiliate and the domestic economy of its host developing nation. Backward linkages are those created by the affiliate's use of local inputs, and forward linkages are those created through the use of the affiliate's output in further local production. Most developing nations believe that the strength of these economic linkages directly impacts the dimension of the wedge between the total and private investment return. For this reason, they are less enthusiastic about foreign direct investment that has negligible forward and backward linkages. They view such investments as the exploitation of their low-cost labor, with little or no residual benefit. Certain nations, such as Singapore, refuse to accept foreign direct investment with negligible linkages to its economy.

Examples of investments with inadequate backward economic linkages are the assembly of automobiles from imported parts, formerly known as CKD kits (completely knocked-down), as are the packaging of pharmaceuticals from imported materials. Examples of inadequate forward linkages are the exporting of logs or ores without further processing into plywood or metals. Investments that have both inadequate forward and backward linkages are those that process imported goods for re-export.

Foreign direct investments that promise strong forward and backward linkages to the local economy add great leverage to the investment proposal by the foreign investor because of the perceived effect the developing nation believes it has on the size of the externalities "wedge." Fortunately, the use of local suppliers and/or subcontractors can also make good economic sense for the foreign investor, in addition to providing stronger backward linkages. This can achieve cost advantages and reduce initial investment levels. It can shorten supply lines and achieve better time-to-market, with lower in-country inventory levels. It can also help protect against unfavorable currency-exchange-rate movements between the currencies of the host country and the home country of the investor.

Successfully using local suppliers and/or subcontractors requires close attention to the process of quality assurance. The FDI affiliate may need to assist the supplier or subcontractor in equipment selection and worker skill training. The affiliate may even need its own staff present in the supplier plant, supervising the work being done for it. However, such techniques and practices and are not unusual between global enterprises and their suppliers in their own home markets and, therefore, should be readily transferable to their FDI affiliates. Outsourcing is a rapidly growing trend in the business strategy of major global enterprises.

PATTERNS OF OPERATION

The potential patterns of FDI operation encompass a broad range of possibilities and are, in fact, only limited by the capacity for investor innovation. As noted earlier, in the countries in transition and in developing nations, one size does not fit all, due to the tremendous social, political, and economic differences between these nations. Thus, one investment concept is not universally applicable.

There can be nonequity accords, such as licensing agreements, franchising agreements, and management contracts. There can be minority-owned foreign affiliates, fifty-fifty joint ventures, majority-owned foreign affiliates, and wholly-owned foreign affiliates. The choice between these patterns depends on a broad range of issues, including the objectives of the host country, finance considerations, tax considerations, the intent to exert some form of control, the need for technology or brand-name protection, and the desire to maximize the use of local material and labor inputs.

Nonequity patterns, which unbundle the FDI package, were initially preferred by developing nations as a means of acquiring valuable resources without inheriting the foreign domination of specific domestic-market segments. Certain global enterprises also saw nonequity accords as a means of market entry at significantly lower risk and investment levels. But these nonequity accords also reduced the level of economic rewards achieved, and sometimes required just as much time and effort to support as did an equity investment.

From the point of view of a U.S. foreign direct investor, there are three critical levels of equity sharing in foreign direct investments. The first of these is 19% and less, the second is 20% to 49%, and the third is 50% to 100%. Under U.S. accounting standards, an enterprise that owns less than 20% of an affiliate cannot consolidate the financials of that affiliate with its own financials. It can only report income received from that affiliate in the form of fees, interest payments, and repatriated profits. The enterprise can consolidate its share of the affiliate's profit, from 20% to 49%, whether or not repatriated, but does not consolidate the affiliate's sales revenues nor assets. From 50% to 100%, the global enterprise completely consolidates the sales revenues, assets, and profits of the affiliate, but must then deduct from the consolidation that percentage of the profits due to minority shareholders. Obviously, at 100% ownership, the enterprise completely consolidates the affiliates profits, sales revenues, and assets.

In addition to the accounting issues, the percentage of equity ownership also influences the degree of management control and the degree of confidentially for technology and brand-name products. Given a choice, developing nations would prefer FDI affiliates in which the foreign investor has a significant minority interest. Such a pattern provides a significant flow of investment funds into the developing nation and provides for a transfer of technology and job skills into the developing nation, but still leaves management control of the affiliate in the hands of local investors. The developing nations believe that the majority local investors are more sympathetic to the social and political goals of the developing nation, and thus have the leverage to cause affiliate behavior favorable to the objectives of the host country. In years past, certain developing nations, such as India, mandated local majority control of FDI affiliates for these reasons. When companies such as IBM and Coca-Cola exited India because they would not vest their technology or brand names in companies that they did not control, the Indian majority ownership law was repealed, and both companies have now returned to India.

In actual fact, the belief that local investors are more sympathetic to the goals of the host country is not necessarily true. Investment capital in developing countries is in short supply, and therefore viewed as quite valuable. Those that have it expect immediate high rates of return from that capital, and often lack the patience needed to build a new FDI affiliate from scratch. Their foreign investment partner, being quite conscious of his foreignness, is usually more patient and more interested in accomplishing the goals mutually agreed to in the investment accord with the host country. As one such local investor once said, "I have no interest in being the richest man in the local cemetery."

Even though the U.S. Department of Commerce deems an investment in a foreign enterprise of 10% or more of its equity to indicate an intent on the part of the investor to exert some form of management control, and thus qualify that investment as a foreign direct investment, absolute control is not gained until the investor owns at least 50.1% of the shares. Of course, effective control can be obtained at ownership levels less than 50.1%, based on the ownership distribution of the remaining shares. Often, for the purposes of perception in the developing world, foreign direct investors are persuaded to share the equity in an FDI affiliate on a fifty-fifty basis with local investors. Under U.S. accounting principles, such sharing also permits both parties to consolidate the sales revenues and assets of the affiliate with the financials of their own enterprises, and to also consolidate 50% of the affiliate's earnings. Other than satisfying certain perception objectives this is not a productive exercise, in that unless certain other actions are taken it leaves the responsibility for the management of the business in doubt. In today's world of accelerated business pace, it is not acceptable to leave this matter of ultimate control in doubt and to raise questions as to who is expected to drive the bus.

One solution is to divide the equity shares in equal halves but then to grant the right to two votes, rather than one, to a share held by one of the two parties, and thus create a "golden share."

Basic decisions, such as the sale, merger, or recapitalization of the enterprise could require a super majority, say 70% of all shares, which means both shareholder groups must favor these critical decisions. In foreign direct investment, appearances are important, but these appearances should not be permitted to interfere with or delay the vital process of business decision-making.

Most developing nations also oppose the 100% acquisition of an existing local enterprise by a foreign enterprise. They view this as alienating the local economy without providing offsetting benefits. This is seen as reducing potential competition and increasing the economic domination by the new foreign-owned affiliate. However, if there were to be a minority equity stake taken in an existing local enterprise by a foreign enterprise, this would be viewed much more positively. There are significant advantages to foreign enterprises to include local equity in their FDI affiliates, in that the local investors usually bring with them valuable local knowledge, including local market knowledge, local management skills, and local government contacts. Many of the potential human-resource problems mentioned in an earlier chapter can be significantly diminished by the presence of experienced local investors.

Although there are advantages in local equity, there are other benefits to global enterprises if they enjoy majority ownership of their FDI affiliates. Most importantly, it permits the global enterprise to retain absolute management control while at the same time enjoying responsible local partners who provide enhanced linkages to the local economy and who reduce the foreignness of the affiliate's appearance. This disadvantage is, of course, receiving a smaller than 100% share of the investment rewards when the vast majority of all resources have been provided by the global enterprise, and the time and effort involved in managing the FDI affiliate are probably the same, as in the case of a wholly-owned foreign affiliate. The level of investment should be less, but occasionally the local investors look to the global enterprise for loans or other assistance necessary to fund their share of equity contributions.

For still other reasons, many global enterprises prefer wholly owned foreign affiliates. Included in these are: that such ownership facilitates the concept of centralized management by the parent system and increases the velocity of decision-making, that it avoids the dilution of the investment return for essentially the same amount of parent enterprise time and energy, that it provides greater security for intellectual property and valuable global trade-names, and that it eliminates the pressure from local partners seeking premature financial returns on their scarce local capital. With the broad range of operation patterns available, a global enterprise must highlight the resources it brings to the party in order to achieve in the investment accord the pattern of ownership it most earnestly seeks. Of particular importance to this negotiation are the alternatives available to each party.

The leverages enjoyed by global enterprises in these investment accord negotiations vary widely based on the type of business in which the global enterprise is engaged. In projects that involve the extraction of natural resources, the leverage favors the developing country. Such resources exist in only certain nations, and enterprises wishing to extract them must come to one of those nations. In manufacturing, a project to produce conventional goods for the local market provides low leverage to the potential investor. A project to provide high-tech products for the local market has greater leverage, and projects to produce high-tech products for the local market and for export have the best leverage. Similarly, projects to provide conventional services for the local market carry low leverage, projects to produce advanced services for the local market have a higher leverage, and projects to produce advanced services for the local market and for the surrounding region have the best leverage. The degree of leverage also influences what FDI incentives may be offered. And, of course, to the extent that the investment proposed establishes strong forward and backward economic linkages to the local economy, that it transfers technology and upgrades local job skills, its leverage will be further enhanced.

IMPORT SUBSTITUTION

An early economic strategy practiced by many developing nations was that of import substitution. Its theory was that balance-of-payments benefits could be achieved by transferring the production of necessary economic inputs into a country to substitute for the former importation of these same necessities and at the same time create additional domestic employment and a new source for exports.

The strategy of import substitution was supported by a series of economic policy instruments. Barriers to the continued importation of the substituted product were created using quotas and/or steep tariffs. Other import restraints used were prior-deposit requirements, foreign-exchange allocations, and multiple currency-exchange-rate schemes. Probably the most stark example was Brazil's Law of Similars, which fortunately is no longer in effect, but which effectively said that no product currently manufactured in Brazil could be imported into Brazil.

Import substitution measures also included mandated minimum, and often escalating, local value-added content requirements, enforced through the threat of price controls, loss of access to government contracts, loss of access to import licenses for parts, and threatened reductions in the level of local access barriers. Minimum levels of export by these protected producers were achieved through the promise of export subsidies. Of course, many of these measures have now been banned by the WTO and the TRIMS Agreement prohibitions on minimum, local value-added contents and export subsidies as a condition for foreign direct investment.

The strategy did provide some improvement in current account balances, transfer some technology, and create some new job opportunities. But at what price? The domestic producers of the protected products raised their prices to a level just below what an imported equivalent would cost, and the high required levels of local content often reduced product-quality levels. Too many deals were struck with too many companies in too many product areas, with no requirement for either price or quality improvement in the future. The local consumers were then stuck with higher-priced, lower-quality equivalents to what they had previously been able to import. The strategy of import substitution often strengthened local linkages to the domestic economy by forcing the use of local inputs to meet the mandated, minimum, local value-added requirements. But the combination of higher prices and lower quality levels made it virtually impossible to export these products without export subsidies.

Unfortunately, many Latin American nations were lured by this "import substitution siren song," and their economic development fell seriously behind that of the nations of the Pacific Rim, which opted for economic development via the manufacture of products destined for export to competitive world markets. This strategy of export orientation involves minimum import duties into the nation for parts and manufacturing equipment and no minimum, local value-added requirements. It produces globally competitive products exportable to global markets without the need for export subsidies. Fortunately, export orientation has now been adopted by most developing nations, including those in Latin America, but nations in the Pacific Rim which initially followed the policy of export orientation got a big head start in economic development over those which initially chose import substitution.

Time finally caught up with the strategy of import substitution. The strategy depended upon restricted market access. GATT banned import quotas, and successive rounds of GATT trade negotiations significantly reduced bound tariff rates. To achieve investment eligibility, the TRIMS Agreement of the GATT Uruguay Round specifically prohibited minimum local-content and trade-balancing export requirements. And finally, the strategy just didn't work. It added some employment, transferred some technology, and improved some current-account balances, but it failed to produce globally competitive outputs and was defeated by a strategy of export orientation.

But import substitution, like other flawed economic strategies, could reappear. Its basic concept is imbedded in the growing demand for production offsets on purchases of products such as aircraft by government-owned airlines. It also appears in the value-added minimums in the rules of origin common to free-trade areas. Economic-development strategies that promote open-market access and that depend on market economics to produce globally competitive outputs will beat the import substitution strategy every time, and its principles of restricted market access and arbitrary limits on factors of production.

In concluding these three chapters on the role of foreign direct investment in the developing world, it is well to remember that these nations are home to 85% of the world's population, with the highest levels of unmet needs. The markets of the developing world have been growing faster than those of the advanced economies. In the majority of these nations, centrally planned economies are giving way to market-force economies. Today, there are developing nations present in the world's largest free-trade areas and customs unions. Each developing nation is different. One business concept or strategy does not fit all. The three most important developing-nation attributes are political stability, economic stability, and social stability. The developing world is well worth what some perceive to be considerable hassle. It is the global area in which the process of economic development and economic growth is taking place. As the developing world evolves, it has the potential to aid in the conversion of global inputs into outputs for global markets, and to aid in the creation of conspicuous customer solutions. But most of all, it can significantly expand the dimension of the global economic pie.

Chapter 13

Reflections

Reflecting on these twelve chapters, which have explored the subjects of technology, innovation, international trade, and foreign direct investment, I hope the case has been made that their enlightened confluence has been and continues to be the key ingredient to conspicuous customer solutions and to the efficient conversion of global inputs into outputs for global markets. In addition, I hope the case has been made that conspicuous customer solutions and the efficient conversion of global inputs into outputs for global markets have been important to the liberalization and further globalization of the world's economy, the improvement of global welfare through the raising of global living standards, the inclusion of more of the world's population in the benefits of expanded global commerce, and the ability of market economics to encourage the practice of political democracy.

However, in spite of the significant dimensions of the above accomplishments, there is an even larger message in the above, which is the unique nature of the second half of the twentieth century and of the world leaders in that period. In fact, this half century may very well have been the most important era in modern history.

As the new era dawned, global integration of the world stood at historic lows, much of the productive capacity of Europe, the

Soviet Union, and Japan stood in ruins, many millions of people had been killed, and poverty was rampant. It was on to this stage that walked the American and British architects of the post–World War II economy, together with their concepts for new international institutions, policies, and procedures to reintegrate the world politically, socially, and economically.

Fifty years later their foresight has resulted in renewed global economic integration, unprecedented liberalization of individual domestic economies, the broad dissemination of market economics, improved global economic welfare, and high levels of international trade and foreign direct investment. At the same time, the strengthened protection of intellectual property rights will now encourage greater investment in research and development to produce the new technologies and ideas so vital to the process of innovation and global problem-solving. An important message in this fifty-year experience is that economic development is highly dependent on the practice of both market economics and political democracy, and that it is counterproductive to sacrifice political democracy in a misguided attempt to accelerate economic development, a fact not always understood by the countries in transition and developing nations.

In spite of sporadic criticisms, the international institutions of this world have provided stability, assistance, and guidance in troubled times, including the United Nations in its social and political role, the World Bank and the IMF in finance, and GATT and the WTO in the areas of international trade and foreign direct investment. Multilateralism and nondiscrimination have dominated much of the growth in global commerce, even though regionalism has enjoyed significant success in situations such as the European Union and NAFTA. Properly coordinated, both regionalism and multilateralism have the capability to successfully converge in an even fuller integration of the global economy.

The economic development of countries in transition and developing nations continues through an agonizing process that has repeatedly demonstrated that you cannot build strong domestic economies through grants in aid, nor by foreign lending, but that these nations must earn their own way into economic development through the acquisition of resources necessary to the performance of useful activities or the production of useful things. Fortunately, the earlier distrust between developing nations and global enterprises is receding as they learn to maximize the benefits to each party in the process of foreign direct investment. Both sides now exhibit much more pragmatic views as to how to mutually share these benefits and have adopted approaches much more accommodating to the needs of each other. In most such nations, the economic development strategy of export orientation has replaced the earlier, nonproductive strategy of import substitution.

Occasional economic crises in the nations of the developing world have revealed flaws and artificialities in their versions of market economics. The 1997–98 Asian financial crisis demonstrated that much of the "Asian Miracle" was unfortunately based on an artificiality of favoring the producer over the consumer. But, in many cases, such crises have highlighted the existence of such flaws and have hastened the adoption of corrective actions. As President Kim Dae Jung of South Korea so aptly remarked in the spring of 1998, "In every crisis there is both risk and opportunity. We intend to maximize the opportunity."

The last half of the twentieth century witnessed the agonizing years of the cold war. Even though that war ended in the early 1990s with the economic collapse of much of the communist world, we are still not a world completely at peace. How to effectively deal with certain rogue states seems to be a constantly evolving political and military dilemma. However, there is no doubt of the necessity to deny these nations access to weapons of mass destruction and terrorism and at the same time to maintain adequate defenses against their military capabilities. In this regard, the United States needs to rigorously review its policies regarding the imposition of economic sanctions, particularly unilateral sanctions, and the influence these have on U.S. relationships with other friendly nations.

Obviously, the last fifty years have not solved all global economic, social, and political problems, even though much progress has been made. There are many new issues that require resolution in the early years of the next millennium. These include appropriate remedies to the threat that global low-wage workers pose to the security of workers with low skill and education levels in the advanced economies, the protection of the global environment as the level of global economic activity increases, the further harmonization of domestic competition policies as the level of international trade barriers continues to fall, and a sorting out of the ultimate consequences of a virtually unfettered flow of capital around the world.

These are great new challenges, but the number of and complexity of similar problems addressed and solved in the last fifty years should give great confidence that in the spirit of the greater good these issues can also be solved. There are more institutions for solutions today than there were fifty years ago. A greater number of world nations now participate in the activities of these institutions, and there is also greater understanding of the interdependence between the nations of this world. What is important is that these issues be addressed promptly and aggressively. The world can't just be content with what has been accomplished to date. Without forward progress there will be backsliding, in that it will then be the only game in town.

So here's to "globalizing conspicuous customer solutions" and to the ability to efficiently convert global inputs into outputs for global markets. As long as we keep these objectives in mind, the appropriate social, political, and economic policies for this world should be easier to discern.

Index

About the Author

EDMUND B. FITZGERALD is Managing Director of Woodmont Associates in Nashville, TN, and an Adjunct Professor at Vanderbilt University. He is the retired Chairman and CEO of Nortel Networks Corp. of Toronto and former Chairman and CEO of Cutler-Hammer, Inc. of Milwaukee.

ISBN 0-275-96995-9

HARDCOVER BAR CODE

DATE DUE

		JUN 0 8 2002	
			Printed in USA